Pocket Guide to
Technical Presentations and Professional Speaking

Steven B. Zwickel
University of Wisconsin–Madison

William S. Pfeiffer
Southern Polytechnic State University

PEARSON

Prentice Hall

Upper Saddle River, New Jersey
Columbus, Ohio

Library of Congress Cataloging in Publication Data

Zwickel, Steven B.
   Pocket guide to technical presentations and professional speaking / Steven B.
Zwickel William S. Pfeiffer.
      p. cm.
   Includes index.
   ISBN 0-13-152962-5
   1. Communication of technical information. 2. Public speaking. I. Pfeiffer,
William S. II. Title.

T10.5.Z95 2006
808.5'1--dc22

2005006705

Editor: Gary Bauer
Editorial Assistant: Jacqueline Knapke
Production Editor: Kevin Happell
Production Coordination: Tim Flem, PublishWare
Design Coordinator: Diane Ernsberger
Cover Designer: Bryan Huber
Cover art: Digital Vision
Production Manager: Pat Tonneman
Marketing Manager: Ben Leonard

This book was set in Meridien by PublishWare. It was printed and bound by Courier
Stoughton, Inc. The cover was printed by Phoenix Color Corp.

PowerPoint® is a registered trademark of Microsoft Corporation in the United States and/or other
countries. Microsoft products are the intellectual property of Microsoft Corporation and are protected by copyright laws and one or more U.S. and foreign patents and patent applications.

Pearson Education Ltd.
Pearson Education Singapore Pte. Ltd.
Pearson Education Canada, Ltd.
Pearson Education—Japan

Pearson Education Australia Pty. Limited
Pearson Education North Asia Ltd.
Pearson Educación de Mexico, S.A. de C.V.
Pearson Education Malaysia Pte. Ltd.

10 9 8 7 6 5 4 3 2 1
ISBN 0-13-152962-5

Preface

Good speeches should start by giving listeners three main pieces of information:

1. Purpose: What purpose does the speech have?
2. Importance: What importance does the speech have?
3. Plan: What plan will the speech follow?

This formula applies not only to speeches but also to writing—from business proposals to handbooks like this one. So without delay, let's examine the purpose, importance, and plan of *Pocket Guide to Technical Presentations and Professional Speaking.*

Purpose of This Book

This book aims to help four main groups improve their skills in public speaking:

- Students in introductory speech classes that focus on the process of preparing and delivering technical presentations and speeches
- Students in other courses—particularly science, mathematics, engineering, technology, and business courses—that require speeches
- Employees in need of an on-the-job reference guide to giving speeches
- Trainers and facilitators who need a handbook for business seminars on how to give speeches and presentations

Not many people in any of these groups actually look forward to public speaking. Speeches force us outside our usual "comfort zone" of interpersonal, informal communication and into a "risky zone" of communication with larger groups in more formal situations. But because most of us do enjoy speaking informally, one key to good public speaking is to transfer the skills that we already display in less stressful contexts—such as conversations with friends or meetings with colleagues—to the formal contexts covered in this book. *Pocket Guide to Technical Presentations and Professional Speaking* will help you make this transition.

You've probably heard that every obstacle in life can be viewed conversely as an opportunity for learning. This book encourages you to see every speaking event as an opportunity to educate your audience. Public speaking also helps you grow personally and professionally. Just as you learn more about a subject in the process of preparing to write a report or proposal, you also learn more about a subject in the process of preparing to speak about it to others. View each speech as a chance to learn as much during the preparation process as you want your audience to learn during its delivery.

In short, improving your speaking skills first requires you to focus your attention and discard negative thoughts. Then you can proceed to follow the guidelines in this book for preparing and delivering your best speech. Adopting a positive attitude and working hard go hand in hand.

Importance of This Book to You

We've established that the act of preparing and delivering speeches requires a great attitude and hard work. It's only natural for you to ask, "Why are speeches worth all this effort?"

The answer is that success in your professional life may well depend on your speaking skills—either because you regularly will be asked to speak before groups or because you will occasionally be asked to do so. In the first case, much of your job may involve presenting new ideas to colleagues in your own organization or speaking about products or services to your customers. In the second case, you may toil away for years before you get that first request to speak before supervisors, clients, or professional colleagues. In either scenario, your next promotion, your next job, or your professional credibility may depend on the skills emphasized in this book.

Besides helping you influence others, there's another, equally important reason to use this book in refining your speaking skills. Every time you deliver a competent presentation you add to a reservoir of self-confidence that spills over into the rest of your life. Success in public speaking breeds success in interpersonal communication because similar skills are at work. Your ability to inform, persuade, educate, and entertain in an oral presentation increases the likelihood that you will be more effective at running meetings, do better in job interviews, and respond more appropriately in a performance evaluation.

Thus your ability to speak effectively in public will enhance the quality of your personal and professional life. Now, how is this little book designed to help you quickly achieve the goal of effective public speaking?

Plan of This Book

The principle that drives this book is that just about anyone can learn to give a competent oral presentation if provided with simple guidelines to follow. The guidelines are derived from the authors' many years of experience teaching communication skills at the university level and in the business world. It aims to give you enough guidance to help you meet the five main challenges of presenters: research, organization, text, graphics, and delivery.

Each chapter presents guidelines, supporting examples, and practical advice. Absent are long explanations and complex theory, which help little with the work of preparing and delivering a speech.

Organization of This Pocket Guide

This guide is organized to help you follow a process for creating and delivering an effective presentation. At the end of each chapter you will find some encouraging words and exercises to help you apply the principles described in that chapter. The Appendix includes a list of resources for public speakers. Here is what you will find in each chapter:

Chapter 1 Overview of Technical Presentations and Public Speaking

- Why speaking is just one part of a much larger communications process
- What the different types of communication modes are
- How to choose the most appropriate communication mode

Chapter 2 Audience Analysis

- How to do an audience analysis to find out what your listeners need and expect
- What you can do to make yourself more credible
- How to be an ethical presenter

Chapter 3 Know Your Purpose

- How to refine your purpose for speaking
- How purpose and method of delivery determine the type of speech
- Where to find credible resources and how to use information and ideas from other sources

Chapter 4 Coping with Anxiety

- Why people feel anxious and stressed when they have to speak in public
- How to cope with anxiety and stage fright

Chapter 5 Organizing Your Presentation

- How to organize a presentation, starting with an outline following the Introduction-Body-Conclusion (IBC) format
- Ways to create a set of presenter's notes
- How to practice your speech before you deliver it

Chapter 6 Illustrating Your Presentation
- How to make visual aids to illustrate your talk
- Ways of using numbers in presentations
- When to use the different types of visuals: text, data graphics, and representational graphics

Chapter 7 Using Graphics
- How to use visual aids when you present
- Why you should interact with your visual aids
- How to get more out of presentation graphics software

Chapter 8 Delivering Your Presentation
- Why both verbal and nonverbal communication are important in delivering a speech
- How to improve your voice and your delivery
- Ways of adding appropriate humor to a talk
- How to handle question-and-answer sessions

Chapter 9 Evaluating Presentations
- Why getting feedback on your speech is important
- How to use evaluation tools to improve your oral communication skills

Chapter 10 Adapting to Different Situations
- How to organize and deliver presentations in specific situations—on the job, giving persuasive talks, doing training, and speaking at public meetings
- Ways of using audio- and video-teleconferencing to give presentations

Appendix: Resources for Presenters
- Books, articles, and websites that presenters will find useful

Acknowledgments

Steven Bernard Zwickel—This project would never have happened without the persistence and encouragement of Gary Bauer, my editor at Prentice Hall, who "handled" me with grace and charm. I am also indebted to my colleagues who teach in the Technical Communication Program at the University of Wisconsin–Madison, especially Sandra Courter, who took a chance and brought me into the College of Engineering, and Evelyn Malkus, who mentored me and introduced me to Sandy Pfeiffer's excellent textbook. I must also thank Claudyne Wilder for her advice and encouragement, and Bruce Hadburg and Eric Urtes for their tales "from the trenches." I would also like to thank Lynne

Cook from the University of North Texas and Suzanne Karberg of Purdue University for their insightful comments on the manuscript.

Most important in keeping me going through the writing process has been my darling wife, Marjorie. Her unwavering support was crucial and her sound advice was indispensable in making this book a reality. Marjorie offered valuable insights into the relationship between speaker and audience and her willingness to review the manuscript in its entirety more than once was truly heroic. She is my hero.

William Sanborn Pfeiffer—I would like to thank our editor at Prentice Hall, Gary Bauer, for suggesting that this book be developed, based on my earlier *Pocket Guide to Public Speaking* (Prentice Hall, 2002). Additional thanks go to Kevin Happell at Prentice Hall who helped get this book through production successfully.

I also want to express my deep appreciation to the following individuals who contributed ideas or text to the original book: Dick Hahn, Dory Ingram, Kim Meyer, Hattie Schumaker, Betty Seabolt, James Stephens, Shawn Tonner, and Stephen Vincent. Some of their material has been brought forward into this new book.

Last, I would like to thank my wife Evelyn, daughter Katie, and son Zachary, for their constant support on writing projects and for their help with the text.

Brief Contents

Contents

CHAPTER **6**

Illustrating Your Presentation   113

CHAPTER **7**

Using Graphics   159

Introduction

THE WAY IT WAS

If you had attended a college class any time before World War II, you would have found yourself sitting in a large lecture hall while the professor, dressed in business attire and standing behind a lectern[1] at the front of the room, read directly from notes for the entire class period. There were no pictures projected on a screen—in most cases there would be no illustrations of any kind. The use of humor was rare, and rarer still was discussion. There was little give and take between the professor and students. And, if they did talk, it was all extremely formal.

Fast-forward to the early twenty-first century and you will find a very different kind of college class with a far more relaxed atmosphere. Changes in the nature of public speaking over the past twenty-five years parallel changes in the way people are educated and entertained.

What Changed?

Everything. The biggest change has been in the relationship between the speaker and the audience. Where public speaking textbooks of fifty years ago stressed the importance of the speaker, the emphasis now is on the audience. It is the speaker's job to analyze that audience and find ways of meeting its needs. As one professor put it, "Don't speak because you have something to say; speak because your audience needs to hear your message." Because of these changes, old-fashioned forms of public speaking, especially the formal speech, are fading fast.

---

[1]A *lectern* is a tall table with a slanted top behind which a speaker may stand. The raised platform on which the speaker stands is called a *podium* or *dais*.

New Kind of Audience

The educational system has become more responsive to diversity and to a much wider range of interests and abilities. In the years since World War II, many different kinds of students with different sets of expectations have begun attending college. The result has been a new kind of audience for lectures. Students today expect their teachers to be accessible, a concept that requires replacing the one-way monologue lecture with a two-way interactive presentation. Because of shorter attention spans and greater time pressures, they expect information to be delivered quickly and efficiently. Audience members expect presenters to be interesting, engaging, and, to a certain degree, entertaining.

Technology Makes More Possible

Our tools for communicating information have changed dramatically. Electronic media—radio, audiotape, TV, videotape, computers, videoteleconferencing, and, ultimately, the Internet—have changed the way people perceive the world and one another. (One popular author of the 1970s, Dr. Marshall McLuhan, argued that the media had become so important that *they* were the message.) These media have sped up communications and increased the importance of visual imagery. Schoolteachers and college professors today have many different ways of delivering their messages to students besides lecturing.

I.2 NEW KINDS OF PRESENTERS AND PRESENTATIONS

There has been a big change in types of public speaking in the past fifty years and a big change in the types of people who need to learn to make effective presentations. Until recently, when we spoke of presentations, we meant "podium speeches" given by high-powered businesspeople and politicians. These were, and still are, formal monologues—one-way communication from the speaker to the audience. CEOs, politicians, clergy, and some university professors still give "podium speeches," but many of them are gradually adopting the more engaging, interactive, entertaining style that is expected of presenters nowadays.

Less Formality

Public speaking is no longer the sole domain of "important" people and celebrities. A much broader range of people today need to learn to make effective presentations—educated people who have ideas and information and who need to communicate clearly and concisely. And many of these presentations are far less formal than the speeches given in the past. Instead of standing in front of an audience and imparting wisdom to the masses, presenters today talk to customers, colleagues, visitors, members of the public, and other professionals.

New Expectations of Presentations

What is required now is a dialogue: a two-way interactive presentation for an impatient audience. Business and technical speakers are expected to give a variety of talks, in addition to the traditional podium speech. Presenters are expected to use visual aids and to provide audience members with handouts. Question-and-answer periods may be extensive and questioners may be probing and, sometimes, hostile.

Shifting Boundaries

At one time, a business speech was something completely different from a scientific presentation. Today, the differences between the professions have become less distinct. There is an increased awareness that the technical fields of engineering, science, and mathematics are connected to the world of business and the boundaries between the disciplines has shifted and become blurred. This has resulted in a change in the kinds of presentations and speeches professionals are called upon to make. For example, engineers may be called upon to explain technical matters to business managers.

The "Technical Presentation"

When we call a talk a "technical presentation," we differentiate it from other kinds of public speaking in that, in a technical presentation, the presenter speaks as an expert on the topic. Thus, technical presenters have a very different kind of relationship to their audiences; they are expected to provide technical expertise.

Compare this technical presentation to the other speeches you may have done so far in your life. When you got up in front of a teacher, professor, teaching assistant, or some other classroom or lab instructor, they were the experts and your job was to demonstrate that you had mastered some part of the material. In those cases, your audience always knew more about your topic than you did.

As a technical presenter, the reverse is true. *You* are now the expert and your audience is less informed on your topic than you are. Thus, a big part of your preparation for a technical presentation is research: learning enough about your topic so that you can handle the material comfortably and professionally.

THE FIRST MODERN-DAY PRESENTER: MICHAEL FARADAY

FIGURE I–1
Michael Faraday: London
(September 22, 1791–August
25, 1867)
(From The Project Gutenberg
eBook, *Beacon Lights of History,
Volume XIV,* by John Lord.)

Michael Faraday discovered the principle of electromagnetic induction and was one of the greatest experimental philosophers of all time. His contributions to science include the fields of chemistry, physics, materials science, and engineering. Faraday was gifted with extraordinary imagination and experimental creativity, and his skills as a lecturer would put most modern-day professors to shame. He made a special effort to share the latest discoveries of the world's scientist with ordinary people, spreading the news of the industrial revolution that was to change forever the way we live.

Here are some of Faraday's own words about lecturing; take them to heart:

> *A lecturer should appear easy and collected, undaunted and unconcerned, his thoughts about him and his mind clear for the contemplation and description of his subject. His action should be slow, easy and natural, consisting principally in changes of the posture of the body, in order to avoid the air of stiffness or sameness that would be otherwise unavoidable.*
>
> *The most prominent requisite to a lecturer, though perhaps not really the most important, is a good delivery; for though to all true philosophers science and nature will have charms innumerable in every dress, yet I am sorry to say that the generality of mankind cannot accompany us one short hour unless the path is strewed with flowers.*[1]

Michael Faraday, in *Advice to Lecturers*

---

[1]*Current Science* 61, no. 12 (December 25, 1991) 787–789.

I.3 *YOU CAN LEARN TO BE A CREDIBLE, COMPETENT PRESENTER*

You've probably heard bad speeches in your life—real clunkers from which you were glad to escape with your life. But no doubt you've also heard some excellent presentations—perhaps at a sports banquet, local lecture series, or political campaign. At the time you may have thought, "What a great speaker!" and wished that you, too, had the gift of public speaking.

Certainly some people take to giving speeches more naturally than do others. However, good speaking is much like good writing in that your ability to follow a disciplined process counts much more than native talent. In its simplest form, the process can be described in terms of the "3Ps":

- Prepare carefully.
- Practice often.
- Perform with enthusiasm.

What never seems to change is the need to prepare carefully before giving a talk. If you learn to use the principles described in this book, you will be ready for anything, including having to make a podium speech.

This book rests on the belief that you can become a good public speaker—perhaps even a *great* one—if you abide by the guidelines presented here. Sticking to them will prepare you to deliver a speech at school, on the job, or in your community.

1 Overview of Technical Presentations and Public Speaking

This chapter includes background information about public speaking. It will describe speaking as one part of a much larger communications process and help you understand that choosing the best communication mode depends on your audience and your purpose.

1.1 SPEAKING AS ONE PART OF A MUCH LARGER COMMUNICATION PROCESS

There are two main parts to the human communication process: Either you are sending a message or you are receiving a message. Thus, depending on whether you are communicating orally or in writing, your role can be described in one of four ways:

Sending	Receiving
1. Speaking	2. Listening
3. Writing	4. Reading

Unlike writing, oral communication is rarely a one-way process. Rather, even as you send a message, your audience is sending messages back to you. This idea—that oral communication is two-way—explains why you need to be "reading" the audience while you are speaking.[1]

To complicate things, an oral presentation may be based on a document, which your audience may or may not have read. For example, consider what happens when people in the academic world present papers (monographs) at conferences. While most technical writers produce a document first, then summarize it in an abstract, academics sometimes write the abstract first and then, after it has been accepted, write a paper to be presented at a conference. This makes it extremely hard to know just what your audience knows.

---

[1] Writers do get feedback from readers, but there is a time lag between reading and response so that it is less direct and immediate than the connection between speakers and listeners.

Another example of how both written and oral communications can complicate a presenter's task is illustrated by what happens when technical experts present written proposals to clients. In these presentations, the experts must pinpoint your audience's needs and concerns or they won't make a sale. And, although audience members have copies of the written proposal, the experts can't assume that the audience has read or understood the document before the presentation.

1.2 CHOOSING THE BEST COMMUNICATION MODE

The three keys to successful public speaking are to choose the right communication mode, to analyze your audience, and to define your purpose. This chapter discusses the first key—choosing the right communication mode. Chapter 2 describes how to conduct an audience analysis so you can prepare and deliver your message to a particular audience. Chapter 3 explains how to define your purpose.

These three key elements of a successful presentation are interrelated. How you define your purpose and the type of audience you want to address will determine the mode you choose.

How do you decide which mode to choose? The following sections provide some things to consider.

Written or Oral Mode

Put your message *in writing* when:

- You want to keep an exact record of your message. If you put something in writing, it may become part of a legally binding contract.
- You want recipients to have time to consider your message and when you want them to have a copy to refer to later.
- Your message is complex or long.
- You want recipients to be able to see your sources cited.

Give your message *orally* when:

- You need immediate feedback from recipients.
- You want to make sure that you don't leave recipients with unanswered questions.
- The subject matter is serious, personal, and may have an emotional impact on recipients to which you want to respond.

MODES OF COMMUNICATING

Modern businesses have many modes for communicating, so you have more choices, which means you also have more opportunities to offend or upset people if you pick the "wrong" mode. Here is a list of the different modes:

Oral Mode

- Face-to-face
- Telephone
- Teleconference

Written Mode

- E-mail
- Intranet posting[2]
- Memo
- Letter
- Formal Document (Proposal or Report)
- Informal Document (Proposal or Report)
- Website

---

[2]An Intranet is a computer network within an organization that is not accessible to outsiders.

Level of Formality

Formal announcements (of promotions, retirements, transfers, etc.) should be made first in person (and in private) and then confirmed in writing. While it is sometimes considered inappropriate to break bad news or to discuss serious personal problems in writing, it may be necessary to follow up with a written memo recording what was said and agreed to. A lot depends on how serious the subject is. Serious subjects deserve the personal interaction of oral communication.

Generally, try to avoid embarrassing people; don't surprise them (even with what you think is good news) unless you are sure that it will be well received. If you decide to make an oral announcement, plan what you are going to say and how you are going to say it ahead of time.

Size of Audience

Because taking up people's time is a problem in the workplace, think about who really must be there to hear what you have to say. Before you decide to call a meeting, consider who has to be in your audience and who might be excused.

If you want to distribute your message to a wider audience, you might want to put it in writing, but also consider teleconferencing, distributing CDs/videotapes, or posting your message (in writing or in the form of a video clip) on a website.

With a small audience, you may find that oral communication is simpler and faster than writing. Here are some examples of how different audience sizes fit different speaking situations:

One-to-one Meeting

- *Job interviews.* Applicant and employer discuss the applicant's qualifications for a position with the employer.
- *Performance reviews.* Supervisor and subordinate discuss past performance and plan for the future.
- *"Elevator (sometimes called "hallway") interview".* Very short, informal meeting between subordinate and superior in which they keep one another up to date.

Group Meetings

- *Team and staff meetings.* Members report on what they are doing and what they plan to do; information from management disseminated; requests for information passed along.
- *Proposal presentation.* An individual or a small team of presenters try to persuade an individual or small group of clients to choose their proposal.

Large Meetings

- *Technical presentations.* An expert presents information to inform, persuade, or instruct audience.
- *Keynote addresses and formal speeches.* A speaker presents personal and professional perspectives to inform, entertain, and motivate audience.

ENCOURAGEMENT

Don't be scared or discouraged if you have to speak in public. You already know how to speak, so it's not like you have to learn from scratch. Build on your strengths, follow the guidelines in this book, and you will succeed in getting your message across.

EXERCISES

1. **Differences between written and oral communication.** How might you handle these situations where Pat, Stan, Luis, and Jeanne work in the same department for Alex, the manager. What would be the best mode of communicating in each situation? Should those involved meet in person or is it better to put something in writing?

   A. Alex has to tell all of them that he can no longer let them make up their own schedules and he has to explain the new procedure for making up schedules.

   B. Jeanne wants to let Alex know that she is concerned about Stan's use of marijuana.

   C. Stan wants Alex to set a policy for personal use of the departmental photocopier.

   D. Alex has to announce that Luis has been given a promotion and a raise.

2. **Planning for an "Elevator Meeting."** Consider what you might say to someone high up in your organization if one day you were to find yourself riding in the same elevator with that person. He or she might be someone in top management or the Dean of your school—someone you'd like to have a favorable impression of you and to remember you. What might you say in 30 seconds if you had this person as a captive audience?

2 Audience Analysis

The second key element in creating a successful presentation (after choosing the right communication mode) is to analyze your audience so that your message is appropriate, the information is accessible, and you meet the expectations of your audience.

The Introduction described how audiences and presentations have changed over the past few decades and Chapter 1 explained choosing the best communication mode, so you already understand the importance of knowing to whom you are speaking. In the first part of this chapter, we will look at audience analysis in terms of:

- Doing an audience analysis and understanding its important
- Gauging the audience in terms of what information to look for, where to find it, what to do with the information you collect
- Analyzing audience members' roles and backgrounds
- Meeting audience expectations
- Understanding the problems and barriers that listeners face

The second part of this chapter discusses establishing the credibility of the presenter—factors that make presenters believable, including a set of guidelines for building credibility and confidence and for establishing expertise. The third (and last) part of this chapter covers the ethics of presentations.

2.1 WHY IS AN AUDIENCE ANALYSIS IMPORTANT?

A man we'll call Bob works as a financial analyst for a large brokerage house. His job is to keep track of corporate stocks in a particular business, which is referred to as "following" that industry. Bob's specialty is following computer companies. If he thinks a company is well run and could be successful, he will recommend that investors buy shares in that company.

As part of his job, Bob travels all over the country visiting different companies and learning more about what they do and how they do it. When he gets there, he is given a folder full of printed materials explaining the company's finances. Then, he usually gets a tour of the facility before he is led into a conference room, where the staff—engineers, technical experts, and managers—give him presentations about the company. For an hour or more, he will be told how excited they are about the future of their organization and why Bob should think it is a good investment opportunity.

There is a lot riding on Bob's opinion. If he gives the company a favorable rating, investors will buy stock and the value of the company will go up. If Bob is not impressed, the company may very well fail. Interestingly, Bob is not a computer engineer; in fact, he has a master's degree in business with a major in marketing. If the company's presentation is too technical—if they use too much jargon and don't explain what they are doing in clear, simple, business terminology—Bob may very well come away with a negative impression.

He must be able to understand what the company is trying to do so that he can explain it to investors. The company presenters must understand their audience—Bob—or their presentation will fail and the company will suffer. That's why audience analysis is so crucial to being a good presenter.

Gauging the Audience—What to Look for, Where to Find It, and What to Do with the Information You Collect

Analyzing your audience can be a challenge, so begin by gathering the kind of information about listeners that will help you design your speech. Write down your thoughts using Figure 2–1, the Audience Analysis Form, to help analyze an audience before doing a presentation. Each of these items will affect how you organize and present your speech. Depending on the type of audience, some factors will be more important than others.

Once you realize your listeners come from varied technical and decision-making levels, you can begin to gather research that leads to a complete analysis of their needs. First, you need to know what to look for, then where to get information about your audience, and finally, what to do with the information you collect.

Guidelines for Doing an Audience Analysis

The next section will help you analyze audience members' personal and professional needs, interests, values, attitudes, and limitations. Use the Audience Analysis Form (Figure 2–1) and the six guidelines that follow to work your way through the process.

What to Look for and Where to Get Information About Your Audience

Group Listeners by Their Backgrounds
Before preparing your speech, determine just how much diversity your listeners encompass. Make a list of either all of your listeners (for a small audience) or a sample of your listeners (for a larger audience). Try to identify listeners by

What I know about my audience:

1. **Their technical roles.** Most of the members of my audience are:

   managers   experts   operators   general listeners

   The audience members who are most important to the success of my talk are:

   managers   experts   operators   general listeners

2. **Educational background.** How much most audience members know about my topic:

   very little   the basics   quite a bit, but not the latest   a lot, including the latest

   Technical Terms: How much of a technical vocabulary do they have?

   Start a list of technical terms that you will use in your presentation. Put a check next to the ones that you don't think you will need to explain. Prepare short definitions of the rest.

   ❑ _____

   ❑ _____

   ❑ _____

3. **Decision-making level.** Most of the members of my audience are:

   decision makers   advisers   receivers

   The audience members who are most important to the success of my talk are:

   decision makers   advisers   receivers

4. **Their interest in the topic.** Most of the members of my audience are:

   very interested   somewhat interested   not very interested   not at all interested

5. **Their needs.** The main question that audience members need answered is: _____

   _____

6. **General preferences in speeches.** Most members of my audience prefer a presentation that is:

   very short   moderate length   long (many points
    (just an overview)   (main points covered)   covered in detail)

   not accompanied   some use of   all major points
    by visuals   visual aids   illustrated

   The attitude of most members of my audience about me is: _____

   _____

   The attitude of most members of my audience about my organization is: _____

   _____

   The attitude of most members of my audience about my topic is: _____

   _____

FIGURE 2–1
Audience Analysis Form *(continued on next page)*

7. Other personal and professional information about audience members.

Needs _____

Interests _____

Values _____

Attitudes _____

Limitations _____

8. Things I still need to find out about my audience. _____

_____

_____

FIGURE 2–1 *(continued)*
Audience Analysis Form

their technical levels and educational backgrounds and by their decision-making level. Next, determine their interest in your topic. Finally, examine their needs, preferences, and attitudes as they affect your topic. After completing the list, you will have a good sense of how the range of audience backgrounds will shape the speech.

List Details About the Audience

You can complete this next step for all listeners or for a sample, depending again on the size of the audience. Realistically, most audiences are too large for you to describe the background of each listener. Instead, you can select a sample of those to whom you will be speaking.

Audience members, even those who are part of a single organization or profession, can differ greatly in their personal and professional interests, values, attitudes, and limitations. Many of the same issues raised in analyzing audience needs are also part of meeting their interests and values. Professional interests may involve meeting organizational goals and fulfilling the organization's mission. Personal interests may focus on opportunities for promotion and increased income.

Values and attitudes are important for presenters to understand because they will determine how the audience responds to the speech. We all filter what we see and hear through our own value systems and personal points of view. Understanding what is important to an audience will make it easier to make sure your message gets through to them.

Finally, a presenter should be aware of audience members' limitations. In an age in which there are so many demands on people's time, energy, and money, it is important to pay attention to how much you can ask of an audience. Generally speaking, the easier you make it for people to do something, the more likely they are to do it. Changing the way people feel and act is particularly difficult. A presenter who expects too much of an audience is likely to be disappointed in the response.

The next two Audience Analysis Guidelines deal with where you can go to get more information about your audience.

Talk with Others Who Have Spoken to the Same Group

The previous two guidelines tell you *what* to search for in your audience research. This one and the next tell you *how* to find the information.

When you accept an invitation to speak, it's a good idea to ask for the name of someone in the organization sponsoring your presentation to whom you can talk about your audience. Whenever possible, try to talk with a presenter who has spoken to the same audience and ask if he or she can give you a sense of who the audience members are and what they need in a presentation. Often your best source is a work colleague. Ask around the organization to find out who else may have spoken to the same or a similar audience. Useful information could be as close as the next office.

Note: This approach only works if the previous presenter has been tuned in to the audience. It's shocking to see how many presenters have trouble answering this question because they never did an audience analysis themselves and because they didn't do any evaluation after the presentation.

Locate Information on the Web

A lot of information about business and non-profit organizations is available on the Internet. If you are experienced at searching the Web, you probably won't have much trouble finding what you need online. If you are not familiar with Internet search engines, help is available at your public or corporate library where the librarians will be happy to help you. Chapter 10, "Adapting to Different Situations," describes how to do audience analyses for special presentations.

If you're speaking to individuals who work in an organization or belong to a professional association, you can often find biographical information on websites of the company, a professional association, or individual audience members. Such sites may describe educational background, work history, professional interests, and more—all details that prove useful in tailoring a speech to the needs of the audience.

Use the Information You Get About Your Audience

The next two guidelines will help you decide what to do with the information you collect by doing an audience analysis.

Focus on the Needs of Decision Makers

The more details you have about all members of the audience, the better you can plan a responsive speech. However, remember that your main focus should be on the needs of those who will act on information in your speech. This subset of the audience should get your special attention in speeches, just as it does in your writing.

In business and industry, audience members are likely to have personal needs that are different from their professional needs. For example, on the personal level, an audience member listening to a presenter discussing environmental issues and proposed laws may want to know more about how his or her home will be affected. On the professional level, business people in the audience may want to know how the proposed laws will affect their way of doing business.[1]

Remember That Most (But Not All) Listeners Prefer Simplicity

Even if you cannot find much information about the background of individual listeners, you can return to a basic premise for all speeches: Many listeners prefer short, easy-to-understand presentations. The popular KISS principle (Keep it Short and Simple) applies to speeches just as it does to writing. Students sometimes feel that the more information they can pack into a presentation, the more likely they will be to receive a high grade, but this is not necessarily true. In fact, many professors prefer depth of understanding to breadth, and a simple presentation that showcases a student's mastery of a topic will earn a higher grade.

However, while few, if any, listeners ever complain because a speech is too easy to understand or too short, audience members who are experts, well educated, intelligent, or curious about the topic and those who have paid a lot of money to see a presentation do not want it to be too easy or too short. It is a waste of their time and insulting to them. Plus, it lets presenters off too easily if they think they can get by with just "kissing."

---

[1] Audience members need different things from presentations. Perhaps you have seen the "Hierarchy of Needs," a list developed by Dr. Abraham Maslow (1908–1970) to explain human behavior. Maslow identified several levels of human needs, the most basic of which, such as hunger and thirst, must be satisfied before the next levels can be fulfilled. Higher levels include needs for affiliation, safety, and love. The highest need is for the fulfillment of one's unique potential, which Maslow called "self-actualization." Depending on how their needs are being met, audience members may have needs at different levels of the hierarchy. Those who are concerned with finding or keeping jobs will be at a different level than those who want professional development. So, some will need basic reassurance, others will want to feel that they are a valued part of—an organization, others need an opportunity to hear what an expert has to say about a topic, and still others need to have the latest information.

2.2   *ANALYZING AUDIENCE MEMBERS' ROLES*

Technical Role and Educational Background

Those listening to a speech probably differ among themselves in technical knowledge of the subject. Although the word *technical* often refers to such disciplines as engineering and technology, it has a broader meaning here. The most "technical" listeners are those who by training, experience, or interest have command of the technical information about a speech topic. Using the criterion of technical role, we can separate listeners into four categories: managers, experts, operators, and general listeners.

Managers

Many employees with technical experience aspire to become managers. Once in management, however, they often become removed from the hands-on technical aspects of their professions. Instead, they manage people, set budgets, and make decisions of all kinds, so managers might not be familiar with technical features of the subject.

Managers often need the following in a speech:

- Background information
- Definitions of technical terms
- Explanations of time frames and budget items
- Grouping of similar points
- Clear statements of what should happen next

Experts

Anyone with a solid technical understanding of the topic may be considered an expert. These listeners may have many years of higher education—as with engineers and scientists—but that is not necessarily the case. For example, a custodial supervisor with no college experience would be considered an "expert" listener for a speech that recommends changes in the maintenance program for a company's physical plant.

Experts may be invited to attend a presentation to offer their opinions about the subject matter and to advise managers on a course of action. Most experts need the following in a speech:

- Thorough explanations of technical implications of the topic
- Visual aids that include relevant supporting data
- References to sources used for technical support

Operators

Managers and experts often make many decisions based on what they hear in a speech. However, an audience also may include operators, who will be responsible for implementing what is proposed and for putting the ideas in the speech into practice.

Operators expect the following:

- Clear organization of the material
- Definitions of technical terms
- Clear connections between the speech content and their jobs

General Listeners

General listeners (sometimes called *laypersons*) sometimes possess the least amount of technical knowledge about, and the greatest emotional investment in, the topic. For example, homeowners may be part of the audience for a speech about the environmental impact of a proposed landfill. They might have little or no biochemical knowledge of the toxic wastes associated with landfills, but they may have a strong interest in how those wastes will affect them personally.

These general listeners often need the following in a speech:

- Clear definitions of technical terms and processes
- Frequent use of graphics
- Distinction between facts and opinions
- References to the effect the speech has on their lives

As you see, there is some overlap among the expectations of all four categories. Yet there are also some major differences that suggest the need for careful research into the technical backgrounds of listeners.

Educational Background

The audience's general educational level may determine the level of sophistication of the presentation. Regardless of which technical category they belong to, listeners may be extremely well educated, but not familiar with the topic of your talk. Even those who know about the topic may not be aware of the latest developments in the field. Don't assume that any audience knows as much as you do about your topic: be prepared to offer short definitions and explanations to bring everyone up to speed and up to date.

One issue presenters must resolve is how to reach a broad variety of people. A presentation for a well-educated group of people might involve more complex ideas and an extensive vocabulary. To help people understand your topic, use metaphors and analogies; add graphics; and have short, simple definitions ready in case audience members appear puzzled or confused. For example, the "right-hand rule" is an analogy used to explain how magnetic fields are created when an electric current moves through a wire.[2] An example of a metaphor might be comparing how information flows through a fiber-optic system to the way water runs through a length of pipe.

Part of your preparation is to discover the educational background and knowledge base of an audience. You can start by asking the person who invites you to speak to describe the audience. If you will be addressing members of an organization, do some research to find out the organization's mission and membership.

You may also do a quick poll of the audience at the outset of your presentation. Ask a series of questions about your topic (be careful not to sound patronizing or arrogant), and have audience members hold up their hands if they know the answers. This tactic will give you a better picture of what they already know and will help you establish the best level for your presentation.

Your job as a presenter will be easier if you can answer the following questions ahead of time:

- What does the audience already know about this topic?
- What definitions and processes do they need to know to understand this presentation?
- What analogies are available because of educational similarities?
- What terms will have to be defined?
- How can visuals be used to help the audience understand the topic?

Decision-making Level

Besides being categorized by technical role and educational background, audience members can be classified by the degree of decision-making authority they have. Pay special attention to decision makers—that is, the listeners who are most likely to create change as a result of hearing your presentation.

Consider the following three levels as you research the decision-making responsibility of your listeners.

Decision Makers

These audience members have the authority to act on the information you present. If you are proposing a new computer lab for your office, for example, decision makers in your organization will decide whether to accept or reject the idea. Or, if you are proposing that students be given an opportunity

---

[2]The right-hand rule is used to show the relationship between the direction of an electric current and the magnetic poles. Supposing that a coil of wire is an electromagnet, if you wrap the fingers of your right hand around it, they will point in the direction in which the current will flow and your thumb will point toward the magnet's north pole. This right-hand rule only works when current flows from a positive terminal to a negative one.

THE TREE OF KNOWLEDGE

One way of finding the appropriate level for your presentation is to think of your topic as part of a "tree of knowledge" (Figure 2–2). Imagine that the broadest classification is the trunk of the tree. As you move upward and out along the limbs, the topics become narrower and more refined until you reach the most narrow sub-subspecialty.

The figure shows that within the broad topic of *biology*, there is a subtopic called *neurobiology*, within which there is a field called the study of *neurons*, which has a subcategory called *axons*, of which one type is *unmyelinated axons*. When addressing an audience that has only a general knowledge of a topic—biology, for example—you would present at a different level than with colleagues in a research lab, with whom you may be able to converse in the technical language of your specialty—the language of unmyelinated axons.

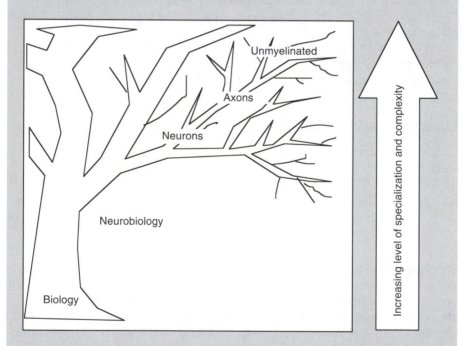

FIGURE 2–2
On the tree of knowledge, the further you go from the trunk of the tree, the more complicated and specialized the subject becomes. Doing an audience analysis will help you determine the right level at which to present information.

to evaluate faculty, decision-maker listeners will decide on the practicality of your proposal. In other words, decision makers have the power to translate information into action.

Advisers

Sometimes the most important listeners may be advisers to whom decision makers have delegated considerable authority. Although they may not make decisions themselves, these are the people who give advice to decision makers on the basis of hearing a speech (and reading the accompanying documentation). For example, those listening to a speech on a proposal to upgrade a company's computer system may include managers who will make the decision and information technology staff who will advise them about the feasibility of implementing the proposal.

Receivers and Implementers

Some listeners are not part of the decision-making process but are crucial for implementing decisions after they have been made. They receive information from observing the presentation and will be expected to adjust their lives and work schedules accordingly. For example, the audience for a speech to the sales department on new procedures for handling Internet sales may include those people who are expected to put the new procedures into practice.

Interest in the Topic

There will be differences between audience members who want to hear you because they are interested in the topic and those who are not interested. Sometimes people attend a presentation because they feel that it is expected of them and not because they have any particular interest in the speaker or the topic. The less interest the audience has, the harder you will have to work to capture and hold their interest. Speaking to an audience that is not attending voluntarily can be a real challenge for any presenter.

As you prepare your presentation, think about your audience in terms of the following questions:

- Why are they coming to listen to you?
- What is their personal or professional interest in your topic?
- Do they expect theories or do they want you to offer them concrete solutions?

The Main Question Audience Members Need Answered

Put yourself in your audience's shoes and try to determine what they most need to know. A good presenter doesn't try to cover too many main points in a single talk. Your main purpose (or purposes) for speaking will form a theme for your presentation. This theme should form the basis for the title and conclusion of your presentation.

2.3 DON'T LOSE SIGHT OF YOUR GOAL

As you conduct your audience analysis, keep in mind the purpose of your talk and consider whether you seek to change your audiences knowledge, behavior, or feelings about the topic. If you want them to do something at the conclusion of your presentation, you must examine the degree to which listeners have the authority to make decisions based on what they hear.

2.4 MEETING AUDIENCE EXPECTATIONS

The importance of doing a serious audience analysis can't be overemphasized. If your presentation fails to achieve its purpose, the most likely reason is that it was not appropriate for the audience. A presenter who doesn't take the audience into consideration will be perceived as arrogant, self-centered, uncaring, and poorly prepared.

General Preferences

The biggest category of complaints people have about presentations is that the presenter failed to meet the audience's expectations. Every speech should aim to satisfy the expectations of the listeners. As obvious as this goal may seem, it remains a deceptively difficult one to achieve. It is fine to be creative, but it is far more important to meet audience expectations.

Format
Over time, audiences for certain types of talks have come to expect particular formats. For example, a math talk is always done on a chalkboard, and archæology lectures are traditionally done with 35mm slide projectors. There are many modern, creative ways of communicating information, but if your audience is not familiar with these and expects something else, you run the risk of alienating some audience members. Be careful!

Level of Complexity
Your audience has a right to know in advance what you will be talking about and at what level. You must present at the right level of complexity to avoid complaints. Most complaints about presentations include the following:

- The material was far too complicated.
- The presentation was much too simple.
- The topic wasn't what was expected.

A CAUTIONARY TALE

A few years ago two engineering students asked one of the authors for help in presenting a design project to the judges in a national competition. We worked over their material so that they had a dynamic, interactive, exciting presentation that we were sure would persuade the judges to give them a high score. When they returned from the competition, they were crestfallen. They had come in twelfth out of fourteen entries. We sat down and did a post-mortem, discovering that the audience analysis was faulty. They had described the judges as being middle-aged men who are interested in science and technology, which we had interpreted to mean that they were "guys like us." When the students described the winning presentation, we realized that this analysis was all wrong. The judges were accustomed to having presenters stand behind a lectern, reading their papers aloud, word for word, without visual aids and without any kind of interactivity. The winning team gave them what they expected, while the enthusiastic, energetic team came in far behind.

These complaints indicate that the presenter had a problem in meeting audience expectations.

Title

What you call your presentation will help set audience expectations, so be very careful in titling your presentation. Puns and wordplay are fun, but try to avoid cute titles and obscure references in a presentation title.

Choose a title that accurately describes the scope and level of sophistication of your talk. If you can't resist the temptation to use a humorous or semi-serious title, add a semicolon and follow it with an informative subtitle that clearly explains what you will be talking about.

> The title "The Bare Bones of Sex" is sure to draw a crowd, but they will be disappointed when they discover that the presentation is about differences between male and female human skeletons.
>
> "Scotch on the Rocks" is cute, but it tells the audience nothing about the presentation. "Scotch on the Rocks: Geological Formations in the Grampian Mountains" is much better and will lead to fewer complaints.

Vocabulary

When you are unsure if your audience understands your presentation:

- Stop frequently to ask for questions.
- Look up to see if audience members are paying attention or if they look confused.

- Prepare "pocket definitions" (these are short, carefully worded explanations that you create in advance and can pull out immediately if an audience member looks puzzled).
- Provide a handout with background information (and a glossary, if necessary) to keep everyone on track.
- Be consistent in your use of technical terms. For example, if you call a part of a machine the "on-off switch," don't refer to it later on as the "power button."

Well-educated people can still have trouble understanding a speech if you use unfamiliar language. For example, a big issue in the field of engineering is that people in different disciplines don't speak the same language and misunderstandings cost companies time and money.

Style

Over time, audience members may have come to expect presenters to adopt a particular style of speaking. For example, motivational speakers are usually energetic, enthusiastic, and focused on personal experience. Thus, they tend to emphasize a few key points that the audience is encouraged to learn and to apply. In contrast, academic paper presentations often consist of having the author read the entire paper to the audience word-for-word with frequent references to source materials and other research. A presenter who tries to do something different or unusual may miss the target and fail to get his or her message across.

It is fine to be creative and to try new ideas, but you must recognize that you run risks whenever you ask an audience to accept an unfamiliar style of presenting.

Length

Time has become more valuable than money, so it is important that you not waste your audience's time. Start promptly. That means that you set up your visuals, arrange your props, test the microphone and amplifier, and adjust the lights *before* you are scheduled to start talking.

If your presentation doesn't have an announced concluding time, tell your audience how long you plan to talk at the beginning of your speech.

Always end on time. (If you really want them to love you, end a few minutes early!)

Time your practice runs so you know how much time your presentation will take. Then plan to leave time for questions, if appropriate. Good time management is a sign of professionalism.[3] Presenters who run over the time

---

[3]See Chapter 4 for an explanation of how controlling your timing will make you feel less anxious and appear more professional.

allotted are perceived as being unprepared, and the audience will resent being inconvenienced.

If you will be talking for more than 75 minutes, build a short restroom break into your speech. If there are audience members who have to leave early, encourage them to pick up handouts and to fill out and drop off evaluation forms before they go.

2.5 UNDERSTANDING THE PROBLEMS AND BARRIERS THAT LISTENERS FACE

Many audience characteristics can prevent a speaker from getting his or her message across. If you are aware of these obstacles to understanding, you can see them as challenges and figure out ways of overcoming them. Following is a list of some of the challenges faced by most speakers. These can get in the way, if you let them.

Listeners Are Often Distracted and Self-interested

Listeners are vulnerable to a variety of distractions. Some of these distractions are external, such as noise from the room next door or problems with the ventilation system in the room where the speech is being given. Others are internal, such as the tendency all of us have to daydream a bit while we're listening to a speech. To keep the audience's attention, a presenter today must be energetic and enthusiastic and the delivery must, to some degree, be entertaining as well as informative.

Listeners Are Impatient

Many listeners lose patience with speeches that are hard to follow, given in a monotonous tone, or last too long. Audience members who are under time constraints will get upset if they think a presenter is wasting their time, so they may ask themselves:

- "What's the point of this speech?"
- "What does this topic have to do with my life or my job?"
- "Where's the speaker going with the topic?"
- "When will this end?"

Indeed, some audience members may have heard so many disappointing presentations that they may not have high expectations for yours. You need to acknowledge to yourself that most listeners are impatient, so that you can prepare a plan to meet their needs. Learn to be concise and clear and to find ways to meet audience expectations so as to justify the time they spend listening to you.

Listeners Lack Your Technical Knowledge

Make information accessible and appropriate to your audience. Define terms and find the right level at which to present. Avoid jargon and insider language, which may confuse audience members. While you want your audience to perceive you as being up to date and modern, if they think of you as trendy, they may also consider you and your ideas as nothing more than a passing fad.

Most Speeches Have Listeners with Diverse Backgrounds

If you could count on the entire audience for a speech always coming from the same background, public speaking would be much easier than it is; your speech could be neatly tailored to the expectations of one like-minded group. Unfortunately, in the real world of public speaking, each audience comprises listeners with different backgrounds, different expectations, and different levels of authority. Diversity becomes an issue in oral communication when the audience includes people from other countries and cultures, when there is a generation gap between presenter and the audience, and when the topic requires a higher level of understanding and literacy.

International Audiences

Business and technology are global enterprises today, and presenters can't count on international audience members being able to understand their jargon, slang, pronunciation, and usage. You will have to work to make information accessible to non-native English speakers. Consider bringing in translators for the hearing impaired and for those for whom English is a second language.

Generational Differences

Talking to an audience that is significantly older or younger than the presenter poses special challenges. With people your own age, you may speak quickly, using slang and frequent references to pop culture. When speaking to an audience that is primarily older or younger, you must switch into a different mode of speaking—one that is more professional. This usually means speaking more slowly, pronouncing words more carefully, and avoiding unfamiliar or inappropriate language.

Be particularly careful when making allusions.[4] If your audience is not familiar with an allusion, you may end up distracting them from your main topic. Different generations are familiar with different allusions. A song, a TV show, or a piece of technology that most everyone in your generation is familiar with may be unfamiliar to audience members of another generation. Your audience analysis should include finding out what age groups you are most likely to encounter in your audience.

Literacy and Depth of Understanding

Literacy is a global problem and most technical and business speakers, in order to reach a wide audience, find that they to have to simplify their language.[5] Almost one-fifth of Americans don't finish high school, and nearly half of the adults in the United States can't read very well.

Learn how to speak slowly and clearly, and to use simpler words, if you want to be understood. Use pictures as a communication tool and be prepared to explain your points two or three times in different ways.

For a presentation to be successful, you must find ways of meeting your audience's expectations in terms of format, level of complexity, and behavior. If you don't meet these expectations, anticipate getting some negative comments on evaluations.

2.6 ESTABLISHING YOUR CREDIBILITY: FACTORS THAT MAKE PRESENTERS BELIEVABLE

Why Should the Audience Believe You?

In an age where everyone seems to be trying to sell something, it is hard for people not to be skeptical about the credibility of public speakers. This is especially true if your purpose is to persuade and if your track record is short. As Professor Evelyn Malkus of the University of Wisconsin pointed out to her undergraduate students, every presenter should think in terms of having two purposes—a main purpose (to inform, persuade, teach, sell, etc.) and a secondary purpose of persuading the audience of the speaker's credibility.[6]

---

[4]You make an allusion when you refer to something without explaining it. For example, you might allude to "the events of 9/11," and your audience will know what you are talking about without your having to explain further.

[5]According to the last National Adult Literacy Survey (NALS) in 1992, 21–23 percent of adults in the United States can read a little, but *not* well enough to fill out an application, read a food label, or read a simple story to a child. Another 25–28 percent of the adult population can perform more complex tasks, such as comparing, contrasting, or integrating pieces of information, but usually *not* higher-level reading and problem-solving skills.

[6]Personal conversation with author.

What Makes a Person Credible?

One of the reasons public speaking is scary is that you know that your audience is watching you from the moment you are introduced and that they are making judgments about you based on your appearance. First impressions are, unfortunately, hard to alter, so it is important that you do everything in your power to come across as competent and believable. Your credibility depends on your being able to present yourself in a professional way.

Guidelines for Building Credibility and Confidence and for Establishing Expertise

Here are some guidelines for making yourself more believable.

Don't Let Anything Get Between Your Message and Your Audience

Don't detract from your credibility by using profanity or offensive terminology. If you want them to hear what you have to say and to take you seriously, don't do or say anything that will distract them or get them *unnecessarily* angry or upset. Be sensitive to their feelings and show respect for their values by not using language that is emotionally "loaded."

Work to Overcome Age Bias

A big issue for young presenters is how to get older audience members to take them seriously. If you look and act the part of the expert, people will be more inclined to believe in you and in your message. It's part of your job as a presenter to let your audience know why they should consider you an expert.

The biggest challenge faced by younger, less-experienced speakers is coping with a major reversal of roles: in school, students present to teachers, profs, etc. who know *more* about the topic than they do. Once they graduate, these roles are reversed. The new graduates take on the role of experts presenting to audience members who know *less* about the topic than they do.

In college speech courses (and even more so as you move on in your career), you will speak to audiences that know less about the topic than you do. Although they may value a technically sophisticated speech, they also want it delivered in language they can understand—you must avoid making them feel foolish or ignorant by "talking over their heads." Think of yourself as an educator. If listeners don't learn from your presentation, you have failed to meet your goal.

As a younger presenter, it may be more difficult to convince an older audience that they should believe what you have to say. There are things you can do and say that will make you appear to be more reliable and trustworthy.

Don't just announce, "I am an expert." Give them specific reasons why: Find ways to point out your education, experience, depth of knowledge, track record, and professional affiliations.

Keep Things Fairly Formal

Use a professional vocabulary and make it a rule to always start a presentation at a formal level—avoid slang and speak in complete, grammatically correct sentences. You can usually loosen up and become less formal if it feels comfortable, but it is very hard to start out at the "Hey you guys . . ." level and then have to become formal.

Make Eye Contact and Use Facial Expressions and Body Language

Remember that nonverbal communication can be more important than the words you use in persuading your audience of your credibility.[7] People will judge your character from your posture and attitude, so stand up straight and look interested and engaged.

Some presenters like to chat with audience members before the talk begins. They feel that this gives them a chance to establish rapport with the audience and see it as another opportunity to find out more about audience members' needs, values, interests, and limitations. If you are not good at making small talk, write out a few remarks and memorize them to help break the ice. Prepare ahead of time and you will be much more comfortable interacting with audience members and responding to questions and comments. Practice speaking slowly and clearly and you will soon adopt a professional-sounding voice, tone, and speaking style.

Make Your Face, Clothes, and Hair Work for You

Appearances count. Dressing like a professional is an important part of establishing your expertise because your audience will decide whether to believe you partly based on how you look. Individuality is precious, but it is extremely important that a presenter look and sound like a professional if he or she is to have any credibility. As a general rule, the more that is at stake, the more formal a presenter's appearance should be.

Guidelines for Appropriate Dress

Some people are apparently unaware of how to dress and when they appear in front of an audience. While many businesses now accept "business casual" as appropriate attire, there are a lot of settings in which formal business clothing is expected of a presenter.

---

[7]See Chapter 6 for more on how nonverbal communications can affect your credibility.

Most professionals should wear "business formal" or "business casual" when scheduled to present. *Note:* Exceptions to these guidelines may be dictated by a presenter's observance of his or her personal religious or cultural practices.

Appropriate Dress for Presenters of Both Sexes[8]

If you are not sure what to wear, follow these guidelines for appropriate dress:

- **Hygiene.** Shower or bathe and brush teeth thoroughly. Use deodorant. Do not use perfume or cologne. Clean and trim your fingernails.
- **Accessories.** Absolutely no piercings. Remove tongue studs. Cover tattoos.
- **Undergarments.** should fit properly and be 100 percent invisible (both color and lines).
- **No hats, caps, bandanas, or do-rags.**
- **A simple watch is OK.** Match watchband with belt buckle color. Necklaces with religious symbols should be worn under your clothing. No large gold chains or necklaces.
- Have a trusted friend or roommate check you over. Do one final check in the mirror!

Business Formal for Men
- Business suit, long-sleeved dress shirt, conservative necktie, dark socks, leather shoes

Business Formal for Women
- Business suit with slacks or skirt (hem below the knee), white or pastel shirt, hose, leather closed-toe shoes with medium heel

Business Casual for Men
- **Jacket and tie are not necessary.**
- **Pants.** Khakis, wool, or cotton blend as long as they are not wrinkled, stained, frayed, or torn. No jeans; no pajamas. Pants should be worn at the waist with a belt that matches shoe color. No underwear should be showing.
- **Shirt.** Short or long sleeve cotton, wool, polyester or cashmere blend with a collar, with shirt tales tucked in. No T-shirts; no crew neck/v-neck, no flannel; no open-mesh shirts. Dress shirts should be 100 percent cotton or cotton/polyester blend as long as they are not wrinkled or stained.

---

[8]Based on advice from U.W.–Madison Engineering Career Services found at *http://ecs.engr. wisc.edu/student/samples/dress.pdf*.

OOPS! WHAT IF YOU MAKE A MISTAKE?

No one is perfect and we all make mistakes from time to time. It is particularly embarrassing to goof in the middle of a presentation.

When you make a mistake in writing, you have the delete key and spellchecking program to make it right. Punch a few buttons and no one will ever know that you wrote six *billion* when you meant to write six *million*.

In a speech, however, there is no way to take back the words that come out of your mouth or the characters that appear on your visuals. So, what can you do if you make an error in your speech?

- If you make a slip of the tongue and your error doesn't really affect the overall meaning of your presentation, let it go. Very few people will notice and it won't make much difference.
- If you misspeak and realize that what you said is substantially different from what you meant and that it will affect what your audience takes away from your speech, then you have an ethical obligation to stop and correct yourself. Don't dwell on your error. Just correct it and keep going.
- The same rules apply to your visuals. Spelling errors are embarrassing, but they are usually not fatal to your credibility. Mathematical errors can be fatal, especially when there is money involved. (Engineers and architects who present faulty calculations are a menace to society.)
- Serious errors that materially change your presentation should be corrected verbally and, if the same error appears in a handout, ask the audience to turn to the page where the error appears and make the correction by hand. If you have not distributed the handouts, determine how difficult it would be (and decide if there is time) for you to correct the error before distributing them.

The important thing isn't that there is a correction to be made, but how you handle the situation. If you treat it as a simple goof (you are, after all, human) and move on, things will go much better than if you keep reminding your audience of how incompetent and clumsy you are.

- **Socks and shoes.** Matching socks that complement pants. Polished and clean dark, leather, slip-on or lace-up shoes are best, but new and clean athletic shoes are OK.

Business Casual for Women
- **A jacket is not necessary.**
- **Skirt or slacks.** Any color is fine as long as it is not wrinkled or stained; white and pastels show up best when presenting in front of an audience.

- **Skirt length.** No shorter than 1″ above knee. If wearing a skirt, practice sitting in the skirt. Choose a longer one if the skirt rides up. No tight skirts or pants! No visible underwear; no visible midriff.
- **Shirt.** Must have sleeves and be opaque; no see-through materials; best choice is 100 percent cotton or cotton/polyester blend in a muted, conservative solid, stripe, or plaid as long as they are not wrinkled or stained. Button-down or straight collar is best. No visible tattoos, cleavage, or bare navels. Avoid scarves or long necklaces that may be distracting or get in the way.
- **Hose and shoes.** Socks or tights should go with your outfit. Polished and clean dark leather slip-on or lace-up shoes are best, but new and clean athletic shoes are OK. Heels: pumps or flats with 1–2″ heel; no spikes. Practice walking in heels before you present.
- **Makeup/nails/hair.** Wear minimal makeup in muted, conservative colors. Avoid long and ornate nails. Trim and clean nails. Solid color or clear nail polish is best. Hair should be pulled back from face so as not to be a distraction. Hair should be clean, freshly cut, combed, and styled, but not overgelled.
- Less is better, so keep accessories to minimum! Simple earrings; no dangles. One earring per ear. Simple ring and watch.

Guidelines for Ethical Presentations

These Guidelines for Ethical Presentations that follow will help you be more credible as a presenter.

1. Do not mislead your audience as to your qualifications, experience, or credentials.
2. Do not mislead your audience as to the topic of your presentation.
3. Do not mislead your audience as to the purpose of your presentation. Don't disguise a sales pitch as an informative talk. Make your purpose clear in the title and at the beginning of your presentation. If your purpose is to persuade, use rational persuasion.[9]
4. Always give credit to sources of the ideas, words, and visual aids you use. Do not "borrow" from others, especially if you are getting paid to present, without arranging with the author or artist for permission to use

---

[9]Rational persuasion involves giving your audience all the relevant evidence and let them make up their own minds. Your job as an expert is to make sure that the information you provide is complete, taken from reliable sources, and up-to-date. Explain why, in your professional opinion, some factors are more important than others. Distinguish between fact and opinion. Demonstrate how you weighed the evidence to reach a particular conclusion you can find this and more information in Robert A. Dahl, *Modern Political Analysis,* 3rd. ed. Englewood Cliffs, NJ: Prentice-Hall, 1976, 25–50.

their materials. The "Fair Use" exception to the copyright law may cover some situations in which infringement might be an issue, but plagiarism is considered a serious offense in *all* academic settings. (See Chapter 3 for more information about using other people's ideas, words, or images.)

5. Know and respect your audience. Don't waste their time. Start on time and finish on time or early. Know why they have come to hear you speak and meet their expectations. Avoid telling jokes or stories that may be offensive.

6. Don't talk down to your audience. Avoid coming across as arrogant or patronizing. If you must use terms—allusions, slang, foreign words, technical terms—that may get in the way of their understanding your topic, define them and make sure the audience understands what you mean.

7. Avoid nonverbal messages that may convey a negative attitude. Just as certain "buzzwords" can make people angry, nonverbal expressions can also upset them.

8. Don't reveal confidential information and don't spread gossip about people or organizations.

9. Answer questions truthfully and completely, including saying "I don't know" when you don't have an answer.

10. Respect other presenters. Don't run over and use up their time. Don't disparage them. If you disagree with what they say, explain why without making it a personal attack.

While most presenters are not bound by a formal code of ethics, the National Speakers Association has adopted a Code of Professional Ethics to which its members are required to subscribe. The principles set forth in the Code (the full text is available at *http://www.nsaspeaker.org*) offer valuable guidance to all presenters. Follow the rules of common courtesy and professional demeanor and you will be accepted as a credible presenter.

ENCOURAGEMENT

The results of an audience analysis can be scary, especially if you discover that the nature of the audience will make achieving your goals difficult. Remember that most listeners really want you to succeed, that most prefer short and simple presentations, and that, if you are not sure if you are reaching them, you can stop and ask if there are questions.

One aspect of connecting with the audience is establishing credibility. You can do this in one way through what you wear. You will rarely go wrong wearing "business casual." But if there is any doubt, ask someone who is an experienced speaker for help. And get used to wearing your presentation outfit *before* you have to present.

AUDIENCE ANALYSIS: TAKING IT TO THE NEXT LEVEL

Audience analysis is so critical to the success of a presentation, especially in persuasive speaking, that it has engendered a new kind of professional consultant. These experts work with businesses to develop marketing plans by identifying customers and pinpointing niche markets.

Attorneys hire these consultants to give them advice about jury selection. Any presenter who needs to know a lot more about an audience should consider hiring a consultant to do scientific research.

Business executives and politicians hire image consultants to help them dress and act in ways that make them credible. A presenter who wants to look fashionable without distracting from his or her credibility can find books, videos, and articles about how to dress for success in the library or bookstore.

EXERCISES

You have been invited to speak. . . . What would you say in the following situations? Analyze each of these audiences and explain how you might adapt your presentation.

1. A group of 13-year-olds who are interested in careers in your field are coming to visit your company. What kinds of things do you think they would find interesting about your field? How would you explain your work to them? *Tip:* Can you remember back to when you were thirteen? What were you interested in?

2. The dean and faculty of your college department are concerned about students plagiarizing and buying research papers online. You have been asked to talk to them about this issue.

3. Several members of a local environmental group will be visiting your place of work and you have been asked to show them around. They are concerned about new laws that may make it harder for your company to comply with environmental standards.

4. A group of state legislators is holding a hearing about a proposed law that would make it more difficult for unqualified people to work in your field. You have been invited to testify.

3 Know Your Purpose

This chapter will explain how to establish the purpose of your speech and how knowing your purpose will help determine the type and structure of your talk. It will help you understand the differences between giving an oral presentation and writing a document. The next part describes the different delivery methods for speeches. The chapter concludes with a discussion of how to collect information for your presentation and appropriate ways of citing your sources when you use borrowed information.

3.1 DEFINE YOUR PURPOSE

The third key element of a successful presentation (after choosing the right communication mode and analyzing your audience) is clarifying your purpose. A lot depends on what you want to happen when you are done speaking. Here are some examples of different purposes:

- **Informative.** Providing the audience with information (facts and opinions)
- **Persuasive.** Presenting professional opinions, usually to change the way the audience thinks, feels, or behaves, and including recommending a particular course of action for the audience to take
- **Occasional.** Entertaining, but includes some informational presentations, usually on a more personal level with humor and some emotional connection to the audience
- **Instructive.** Explaining a process, teaching a skill, or defining terms for an audience, or showing a problem-solving process

Purpose May Determine the Type of Speech

Depending on the purpose of your presentation, you may decide to use a different type of speech. For example, let's say you are a training expert at your firm and assume you've been asked to speak at the firm's annual board meeting about advances in distance education—especially Internet courses. Like many speeches, this one incorporates aspects of all four main purposes for oral presentations: to inform, to persuade, to instruct, and to entertain. You are informing the board about recent advances, you are persuading them you have credibility to speak on the topic, you will be instructing them in how to use the terminology of distance education, and you'd better be entertaining them to some degree because they will be sitting through eight other speeches at that meeting. Despite its multiple purposes, your speech to the board falls mainly into the category of an informative speech.

Types of Speeches Characterized by Purpose

In most speeches one of the following purposes predominates. (See Chapter 5 for more information about formats for presenting professional opinions and showing a problem-solving process).

Informative Speeches

Chapter 2's discussion of audience, divides listeners into several main groups, of which "decision makers" are the most important. They usually want information that will move them closer to making up their minds on an issue. In other words, they may just want information from you, not argument or strong opinion. Many of the speeches that you will be asked to give will take this form. Evidence that you have achieved your informative goal comes from an increase in audience members' understanding of your topic. Examples of informative speeches include introductions to other speakers, keynote speeches at meetings, and kick-off speeches for conferences and seminars.

Informative presentations often require the speaker to explain and expand upon the contents of a written document, such as a report. Your goal as a presenter in those cases is to help your audience understand what is in the document and how it will affect them. You can't assume that everyone in the audience has read the document, nor can you assume that those who have read it understood it or appreciated its importance. In fact, some experts say that if the information in the document is important enough, a presenter should read the entire document word-for-word to the audience. That's why the importance of doing a thorough audience analysis can't be emphasized enough.

Audience members are often pressed for time, so rather than offering them a step-by-step explanation that leads to a broad conclusion (referred to as the Specific-to-General method of development), it may be preferable to start by explaining your conclusion and then going back to show how you collected and weighed the evidence (General-to-Specific). Specific-to-General is more dramatic and a speaker can "hook" the audience's attention the way a good mystery writer presents the clues and names the killer on the last few pages. The General-to-Specific method is lacking in suspense and is comparable to starting a mystery by announcing in the first few pages that "the butler did it" and then explaining how the detective reached that conclusion. It takes longer and it is more exciting, but this method is not appropriate for time-pressed audience members.

Following is an overview of the basic pattern for an informative speech and a brief example of a situation that requires one. Here are the main patterns used to develop the content of an informative speech:

- Definition
- Description
- Classification/division
- Comparison/contrast

The following example shows how a presenter might employ several of these techniques in an informative speech.

> Assume your employer has decided to open an office in Tokyo, Japan. As a member of the human resources staff, you have been asked to give a speech on Japanese culture to employees who will be visiting Japan. These employees will be opening the office, hiring local employees, and meeting with Japanese clients. In presenting highlights of Japanese culture, you might *define* some common Japanese terms, *describe* features of Japanese society, *compare* Japanese and U.S. formalities in writing, speaking, and interpersonal communication, and point out *contrasts* between the two.

3.2   *PERSUASIVE SPEECHES*

Persuading people is one of the most difficult tasks faced by presenters. We are so inundated with persuasive messages and marketing that we have become numb and skeptical of anyone who tries to change our minds about something.[1] Because it is so important, persuasion, rhetoric, and logic have become a major part of every communication course. When preparing to do a persuasive presentation, consider what will influence audience members and think about the ethics of persuasion.

Although many situations require an informative speech, other times listeners expect to hear opinions. In the work world, for example, you may be

---

[1]The scientific study of how people's attitudes change is described by the Social Judgment Theory developed by Carolyn Sherif, Muzafer Sherif, and Roger Nebergall in the 1960s. In a nutshell, the theory says that it is difficult to move people very far from the position at which they start, but it can be done, depending on how strongly they feel about that position. For a discussion of how the theory works, see Daniel J. O'Keefe, *Persuasion: Theory and Research* (Thousand Oaks, CA: Sage, 1990), 179.

the technical expert asked to give your views on a proposed change, such as shifting to a different kind of health maintenance organization or developing a new product. In the academic world, your instructor may ask you to present your solution to a problem or to present one side of a public policy issue. In both cases, your aim is to persuade the audience to accept your views. Evidence that you have succeeded in reaching your persuasive goal will come from new attitudes and behaviors on the part of your audience.

A well-organized persuasive speech is built around a single proposition—a statement of fact, value, or policy that is the heart of the presentation.[2] The process of distilling all your thoughts about your topic down to one main idea that defines what you want your audience to know is called *framing the proposition.*

Once you know the subject of your speech, you can develop a proposition, define the issues you want to address, and collect credible evidence to support your proposition. (Some people call a proposition a *thesis statement.*) It is a single, declarative sentence that defines your main point. The main types of propositions are:

- **Fact.** A disputable assertion that something is, was, or will be
- **Value.** An idea, policy, person, or thing is good (beneficial, moral, justified) or bad (harmful, immoral, unjustified)
- **Policy.** An argument that some action should or should not be taken[3]

The type of proposition will help determine the organization of your speech. Some common techniques for developing the body of a persuasive speech, include:

- Using evidence correctly
- Choosing the most convincing order for points
- Being logical
- Citing only appropriate authorities
- Avoiding logical fallacies
- Refuting opposing arguments

Once you are aware of what you want to communicate, you can use these guidelines to frame the issue in a persuasive presentation:

- Key terms defined
- Issue placed in context and compared to similar issues
- Scope of impact: Number of people affected
- Proof of a logical cause-and-effect relationship
- Reality check: Feasibility of different approaches

---

[2]Barbara L. Breaden, *Speaking to Persuade* (Ft.Worth, TX: Harcourt, Brace College, 1996), 30.
[3]Ibid., 31.

- Validity of sources: What evidence is available and why should it be considered reliable.[4]

The case study that follows would require you to use various persuasive techniques in combination with some informative ones. It involves a context similar to one used earlier for an informative speech.

> Assume your company produces health products for hospitals and is considering whether to open a sales office in Tokyo, Japan. Your sales to Japan have grown steadily for ten years, and you are spending considerable money sending sales representatives to Japan to meet with customers. As a member of the human resources staff, you've been asked to analyze the data and recommend for or against starting the office (Topic).
>
> Having decided to recommend opening the office (Proposition of Value) in your speech, you define the issues and collect evidence to support your proposition. You decide the two issues to address (Framing the Issues) are cost and benefits of hiring native Japanese employees. Thus, you look for facts to help you (1) contrast the cost of sending staff to Japan for the next five years versus the cost of operating an office in Japan for that same period, and (2) give four main reasons for hiring a native Japanese person to run the office, rather than a U.S. employee with experience in Japan.

3.3 OCCASIONAL SPEECHES

Besides speeches that inform and persuade, a third category of speeches is those that engage the attention of listeners at a special event. Of course all speeches must engage the listeners' attention or you risk boring them so they don't hear what is being presented. However, for most occasional speeches entertainment is the main goal. This section gives an overview of some simple techniques to use in such occasional speeches and presents a case study that involves such a speech.

Listeners have high expectations for presentations that aim to entertain, both because they know material will be light and also because these sorts of speeches often occur at the beginning of events when their attention is high. Here are some guidelines for developing any speech that aims to entertain an audience:

1. Emphasize narrative—that is, stories—over explanations. Use anecdotes that help illustrate your main points and practice them so that they sound natural and rehearsed.
2. Use humor with care, making sure to gauge your audience well. Avoid telling jokes, which can easily backfire.

---

[4]Ibid., 55.

THE ENTERTAINMENT COMPONENT

Public speaking has long been a form of entertainment. In the 1800s, speakers like Mark Twain, Charles Dickens, and others were a major draw all over the country. Radio, the movies, and then television changed all that.

As pointed out in Chapter 1, audiences today expect presentations to have an entertainment component. Some speakers mistakenly believe that this means that they have to tell jokes and do a comedy routine. But the dangers of an amateur comic alienating people or failing to amuse the audience are great. Unless you have the ability to tell a joke well—the timing and delivery that distinguish professional comics—stay away from telling jokes. Murphy's Law ("Anything that can go wrong, will go wrong.") says that if you try to tell a joke, you will:

- Forget the joke.
- Remember the joke but forget the punch line.
- Tell the joke wrong.
- Follow a speaker who told the same joke better.
- Offend people unintentionally, thereby obscuring your message.

If you use humor, keep it gentle and make yourself the object of the joke. Anecdotes are much more effective than jokes.[1]

---

[1]An anecdote is a story that helps illuminate your background, that demonstrates your talents, and that offers insight into your motivations and personality.

---

3. Use gestures and other body language liberally. Practice these movements as you practice your speech.
4. Choose the appropriate length. Always stay within the time limits set for your talk and never, ever run over, which takes time away from other speakers.
5. Avoid digressions that can derail an occasional speech. Stay on track so the audience feels that you have used their time wisely.

This example of an occasional speech is similar to that used for the previous examples for informative and persuasive speeches.

Assume that your firm has decided to open an office in Tokyo, now that business prospects in Japan appear to justify such an investment. As the person in charge of the Japanese market for the last five years, you've worked with Japanese customers and have traveled to Japan many times. For a banquet at the company's annual meeting, you have been asked to present an after-dinner speech about your experiences in Japan. Management believes some light remarks will be appropriate, considering the decision to open a Tokyo office. The audience will include middle and top management of the company, corporate directors, and some invited guests from Japan.

Based on what you know about the audience, you've decided to start with an anecdote that illustrates how easy it is to miscommunicate with people from other cultures. You plan to tell a few humorous stories about inadvertent mistakes you have made in adjusting to Japanese culture, in hopes of helping employees who follow you to Japan. Though mainly entertaining, your stories will remind colleagues to bow appropriately, use business cards often, and have their own interpreter rather than relying on the customer's.

3.4 INSTRUCTIVE SPEECHES

As you move up to more challenging jobs, it is likely that training people will be part of what you do. An instructive presentation may include teaching a skill, explaining a process, defining terms for an audience, discussing organizational goals, or training people to follow a troubleshooting process. Good organizations expend a lot of time and money on training, because they understand the value of preparing people to meet organizational goals.

An in-depth look at modern educational theory is beyond the scope of this book, but you should be aware of the fact that people have different ways and different speeds of learning, so what may work well with one trainee may fail completely with another. The mark of an expert trainer is a person who has many different ways of teaching and who can adapt to the situation. If training becomes a major part of your job, you might want to consider taking some professional development courses in educational learning theory. These are available at colleges and universities and through professional organizations, such as the American Society for Training & Development (ASTD).[5]

S-L-M-R ("Slimmer") Speech Model

The S-L-M-R (pronounced "slimmer") speech model presented in Figure 3–1 applies to all forms of speaking. As the acronym suggests, this model is indeed "slimmer" than the complex formulas sometimes used to describe oral

---

[5]An excellent resource for learning more is the American Society for Training & Development (ASTD). They have been around for sixty years and have 70,000 members and associates from more than 100 countries. Contact ASTD at 1640 King Street, Box 1443, Alexandria, Virginia, 22313-2043 U.S.A., phone: (703) 683-8100, Fax: (703) 683-8103, website: *http://www.astd.org/ ASTD.*

FIGURE 3–1
The S-L-M-R ("slimmer") speech
model

communication. It also places more emphasis on the importance of the speaker getting listener feedback, both during and after the speech.

The S-L-M-R model shows that oral communication is two-way—the speaker and the audience are connected in both directions. Rather than speaking being a one-way street, with the speaker sending out information and the audience absorbing it, the speaker must learn to tune in to the audience and to respond to their questions and comments (verbal cues) and to their facial expressions, body language, and voice tone (nonverbal cues). A speaker who disregards these messages from the audience about how they are responding to the presentation may miss the mark and fail to achieve his or her purpose.

Considered together, the four S-L-M-R components—speaker, listener, message, and response—create the context for every conversation and every speech you'll ever present. Following is an overview of each part.

Speaker

As a speaker, you exert the most control over the communication context. Your words have the power to inform, persuade, instruct, or entertain. Your attitude, tone, and general demeanor can sway a listener in many different directions.

Of course, speakers rarely exert total control over a speech event. In the real world, you often lose partial control for several reasons:

- Someone else may have selected the content and purpose of the speech—for example, when a supervisor asks you to speak at a meeting.
- The medium may have features not of your choosing—for example, a conference room may have poor lighting or acoustics.
- The listeners may present you with a challenge—for example, when they have very little knowledge of your topic.

However, these challenges don't change the fact that you, as speaker, must give a successful speech in spite of the obstacles you face.

Listener

Like readers of written communication, when you give an oral presentation, your listeners are your "bosses," and their needs should drive every decision you make in planning and delivering a speech. Remember that members of

DIFFERENCES BETWEEN WRITING AND SPEAKING

One of the most basic concepts beginners must grasp is that there are fundamental differences between writing and speaking. You may write and speak in an informal way with your friends, family, and peers (in a social setting). In a professional setting, however, you can't write the way you speak and you can't speak the way you write. Here are some key differences between written and oral communication:

Writing	Speaking
Words-on-paper is all the reader has; tone and emphasis must be inferred from the text.	In addition to words, listeners have many nonverbal cues, including voice tone, facial expression, and body language.
Writers can go into much greater depth.	Speakers are limited to a few main points; they are rarely able to go into depth.
There is no opportunity to stop and question author while reading.	Speakers often invite questions from audience members who don't understand something.
There is an opportunity to go back as often as reader wants to reread text; it is easy to compare one part of a text with another.	There is no way to "rewind" an oral presentation unless it has been recorded electronically or transcribed.
Readers have as much time as necessary to study illustrations.	Illustrations are available only as long as presenter chooses to leave them up.
Size of page determines size of illustration; fine details may be lost in printing. Whatever information about the illustration is on the page—callouts, captions, sources cited—is all the reader has to go on.	Illustrations may be projected to several hundred times their original size; presenter can explain concepts and discuss details.
Writers usually have no idea if readers understand what they are trying to communicate.	Speakers who are tuned in to how an audience is responding can go back, rephrase, or expand on ideas.

the same audience can differ markedly in their education, professional background, family history, culture, nationality, gender, and expectations for the speech. You may think it's easier to convey information to a listening audience than to a reading audience. After all, there's no doubt that the audience is hearing—though, not necessarily listening to—your speech. With writing, you have no idea if readers have put down your written report after the first page. However, once you recall how your own mind drifts during boring speeches, you will realize that listeners can "check out" just as quickly as readers who put down a document to read their e-mail or go to lunch. (Chapter 2 gives more details about researching the needs of your audience.)

Following are some challenges you face in meeting the needs of your listening audience:

- They have diverse backgrounds and different levels of understanding.
- They may not want to be listening to a speech right now.
- They may have short attention spans or be pressed for time.

Certainly public speaking presents no more important challenge than meeting the needs of listeners. In fact, just as the rule "Write for your readers, not for yourself" drives the modern business writing process, the following rule lies at the heart of your preparation for every speech and presentation: Speak for your listeners, not for yourself.

As obvious as this directive may seem, it has been ignored by many a speaker, resulting in audience boredom, resentment at the speaker for wasting valuable time, and anger at whomever decided to give this speaker a forum.

Message

Every speech involves two-way communication. The part that goes from the speaker to the audience is called the "message," and it has two main parts: content and form. Content includes the information you deliver. Form includes the structure of the message, its style, and the features of your delivery.

Crafting an effective message requires considerable time, which must be used wisely. Many speakers end up spending too little time on the most important tasks (such as outlining and practicing) and far too much time on less important ones (like making fancy visual aids).

Response

Although they understood that the audience may ask questions, some experts fail to place enough emphasis on (a) the reaction of listeners during and after the speech process and (b) the corresponding response of the speaker. That's why the S-L-M-R speech model in Figure 3–1 gives equal treatment

PRODUCING AN EFFECTIVE MESSAGE

Here are some essential tasks related to producing an effective message and where you will find them discussed in this book:

- Locating specific information about your listeners (Ch. 2)
- Shaping the speech around audience expectations (Ch. 3)
- Controlling your excitement and nervousness so that they don't interfere with your presentation (Ch. 4)
- Following a simple structure for the text (Ch. 5)
- Finding and adapting illustrations for your presentation (Ch. 6)
- Practicing in a way that prepares you for the "real thing" (Ch. 7)
- Using your voice and body language to deliver your speech effectively (Ch. 8)
- Staying flexible enough to respond to feedback (Ch. 9)
- Adapting your message to different audiences and situations (Ch. 10)

If these seem like common sense, they are. Like a good piece of writing, a good speech embodies simplicity in design and delivery.

to communication that moves from speaker to listener, on the one hand, and from listener to speaker, on the other hand.

Listeners have different ways of responding to a speaker: through non-verbal (and occasionally verbal) behavior before and during the speech, in questions immediately after the speech, and as questions or comments received later. Two of these types can influence the speech itself, and the third can influence your next communication with the audience. Here are a few points to consider about audience responses:

- Establish rapport with your audience before the speech by greeting them and making small talk
- Observe body language during the speech and adjust your speech in response to the audience. For example: if they look puzzled, ask if anyone has a question. If they look tired, give them a ten-minute break.
- Handle the question period as a continuation of the speech; give a second conclusion when you are done taking questions.
- Provide follow-up as closure. Finish on time and end with a call for action. Let the audience know where they can learn more about the topic.

This approach to understanding and reacting to the response of the audience places great importance on the audience portion of the speech model.

Speech Types Based on Delivery Method

In the first moments of a speech, listeners can observe what method of delivery the speaker has selected. In this instance they form an immediate impression of the speaker and the type of experience they expect to have. If the speaker is reading a text from behind the lectern, for example, his or her speech will be received differently than if the speech were being delivered from a few note cards by a speaker wandering around the room.

Methods of delivery range from the entirely spontaneous impromptu at one extreme to a carefully scripted and memorized speech at the other. These are the four main types of delivery method:

- **Impromptu.** No preparation; no visual aids except for the speaker's own body and any props that are readily at hand
- **Extemporaneous.** Researched results in an outline the presenter can follow (without prescribing the exact words that will be used); visual aids and handouts prepared ahead of time
- **Verbatim.** Speaker reads text word-for-word; may or may not include visual aids
- **Memorized.** Speaker recites script; may or may not have visual aids

The preferred method of delivery for most speeches nowadays is extemporaneous. Let's examine each type of delivery method in greater depth.

Impromptu Speeches

An impromptu speech is one delivered on the spot with no serious preparation. Two colloquial terms used to describe impromptu speaking are "talking off the top of your head" and "working without a net." It can be nerve-wracking to try to deliver a competent presentation on the spur of the moment. Rarely used in formal settings, impromptu speaking occurs when you feel compelled to speak on an issue at a meeting or when others request your opinion or expertise. Here are some guidelines to help you with impromptu speaking:

- **Try to appear confident** (even if you have to fake it). Take a deep breath to steady yourself. Pause to sip some water. Give yourself a moment to prepare. Look like you are thinking hard about your answer.
- **Decide on your conclusion first,** so that the rest of what you say helps you make your point. If you watch the political pundits on television, you'll notice that the best ones start their comments with a flat statement, then they go on to back it up with other ideas until the moderator cuts them off—for example: "The president is really going to have a big fight on his hands over this bill. First, the opposition. . . . Second, he has already stated that. . . . "

- **Start by making a general statement** or by giving background information. This will give you more time to think while you are reciting facts.
- **Begin each of your supporting reasons with the word** *Because* and don't stop until you can stay organized without it—for example, "*Because* of the rising cost of tuition, and *because* of the high price of textbooks, the University is going to have to change its policies . . . ."
- **Answer questions as honestly and directly as you can.** If you don't know an answer, you will have to find some polite way of redirecting your questioner. (See Chapter 8 for suggestions on how to run a question and answer period.)
- **Be very careful not to exceed your authority.** Do not make promises you can't keep and don't bind your organization to a course of action unless you really have the power to do so. Remember that oral promises can become part of binding contracts. When in doubt, tell your audience that you don't have the authority to respond.
- **Finally, tune in to your audience as you speak.** If you are in such a rush to make your points before you forget them, you may be tempted to tune out your listeners. This is a mistake. By watching their expressions and reading their body language, you can find out if they are following your reasoning and whether or not they agree with your conclusions.

Dr. Susan Huxman of Wichita State University suggests that, when speaking impromptu, you should answer questions by:

1. Making a single point,
2. Giving reasons and examples to support your point, and
3. Restating the point.[6]

Above all, don't lose your cool!

Some speakers handle the impromptu format better than others do. They simply perform better when called upon to express an opinion without any preparation. Yet all speakers stand a better chance of delivering good impromptu speeches if they master the techniques of extemporaneous speaking, because most of the delivery techniques are the same.

Extemporaneous Speeches

The term *extemporaneous* refers to a speech that is delivered from notes. Though thoroughly familiar with the material, the speaker does not commit the speech to memory and instead develops final wording during the speech.

---

[6]Cheryl Hamilton, *Successful Public Speaking*, (Belmont, CA: Wadsworth Publishing, 1996), 198–99.

Extemporaneous speeches are by far the most common format used in business and during professional conferences. They are preferred for the following reasons:

- The speaker appears to be more natural and less formal.
- The perceived informality reduces the distance between speaker and audience so the material is more accessible.
- The speech can easily be altered to respond to changing circumstances.

The most powerful argument for choosing extemporaneous delivery is the third point. You communicate best with listeners when your mind stays active and responsive to changing conditions. An audience can sense the sincerity in such a presentation and appreciates the effort it takes.

Verbatim Speeches

In this method of delivery you read a speech word for word, either from a hard copy or from a prompter screen. (Note: The word *verbatim* comes from Latin and means "using exactly the same words.)"

Because few people other than television broadcasters and politicians use prompters, we'll focus on reading from a hard copy. Speakers prefer this mode of delivery:

- When they believe the exact phrasing of the speech is crucial to its effect on the audience
- When they're concerned they may speak too long or wander from the topic if they don't have a prescribed text
- When they feel more comfortable having a prepared text so that they only have to read the words on a page and not "think on their feet"

For most speakers, it's a bad idea to read a speech verbatim. Too often the delivery turns stiff and overly formal with little eye contact with the audience. Feeling ignored by a speaker who drones on, listeners will daydream and finally mentally depart altogether. In fact, some people are so put off by a verbatim speech that they will *physically* depart from the scene. It's not clear that the speaker, with eyes lowered and nose buried deep in the speech text, will even notice.

One reason people are put off by this kind of speech is because they find it demeaning to have someone reading to them like they were children. This is made worse when the text of the speech has been published and audience members have already read it.

A verbatim speech is rarely the right choice for method of delivery, but there are exceptions. One exception might be a lecture by a well-known person, who is also an excellent writer. The speaker might be able to deliver a speech or academic paper while still maintaining eye contact with the audience and creating an animated delivery.

PROMPTERS

In order to appear more spontaneous and engaging when reading a scripted speech, politicians and newscasters use prompters (often referred to by the original brand name TelePrompTers), especially when appearing on TV.

You can often tell when someone is using a prompter—there will be one or two sheets of what look like glass (actually a half-silvered or two-way mirror) angled down between the speaker and the audience or you may notice a black triangular box mounted on the front of the television camera. The sheet of glass reflects the image from the monitor so that the speaker can see and read the text. Because the glass is semitransparent, the light from the scene being videoed can pass through it to the camera lens. The reflected text would not be readable because the mirror reverses the image, so after the text of the speech is typed into a computer it is electronically reversed left-to-right so the image the speaker sees will appear readable. Using two prompters, one on either side, enables a speaker to scan the audience and to look much more polished and relaxed.

A speaker who uses a prompter requires the assistance of another person to scroll down through the text at just the right speed. The assistant must get the timing right so that what appears on the prompter is correct. If you are ever asked to give a speech using a prompter, spend time practicing with the equipment and coordinating the timing with whoever will be scrolling down the text for you.

To understand what it feels like to use a prompter, type a portion of your speech on a word processor, then enlarge the font to 48-point size. Put the cursor on the scroll bar and try to read the text aloud as you scroll down.

Another exception to this is presenting a paper at an academic gathering. For reasons that are not clear, academia clings to the centuries-old tradition of having scholars read papers verbatim in a lecture format. If you are invited to present a paper in this type of setting, do the best you can to make your reading interesting and engaging.

Finally, if your purpose is to promote a book, then reading verbatim from your own work is an appropriate method of delivery.

Memorized Speeches

Like a verbatim speech, a memorized speech usually starts with a text written out word for word. The speaker then practices the speech to the point that he or she commits it to memory and needs no notes or outlines.

Speakers with good memories sometimes choose this format for the following reasons:

- They believe it reduces the likelihood they will wander away from the main topic.
- They think memorizing the speech shows the audience that they have prepared well.
- They believe they will be able to focus more on delivery if they are not dealing with an outline or notes.

Ironically, memorized speeches work best if the speaker looks down at some notes even when they are not needed or used. This technique makes the speaker seem less wooden. However, it's best to avoid memorized speeches altogether. Besides making you seem too stiff and formal, they can go awry (Murphy's Law, again) if you forget part of the speech or lose your place after getting interrupted during delivery.

3.5   CHOOSING A TOPIC

This section applies only to those times when presenters have a choice of topics. Once you enter the workplace, you probably won't have a choice of topics and, in many cases, you won't have a choice of audience, purpose, or setting.

If you have an opportunity to choose a topic for a speech, consider one based on your own experience and training. Following are some guidelines for finding a topic for a speech.

Review Your Life Experience

You speak best about what you know well. Personal experience provides a wealth of information from which you can draw for presentations. Informative, persuasive, and occasional speeches alike can flow from the events of your life. Here are a few general subjects to examine in the process of reaching a narrower topic on which you will speak:

- Skills learned at summer jobs or at work during the school year
- Interesting characters encountered on the job
- Close calls in your life
- Some "first time" experiences that taught you lessons
- Volunteer work that changed your life
- Travel experiences that taught you lessons in multiculturalism
- Characters in your family history, recent or long ago

Consider Hobbies or Special Interests

Good topics are those about which you have personal knowledge. If you look carefully at how you have spent your spare time, you probably will find a topic that will interest others—and you can show why it interested you. The trick in speaking about hobbies is conveying to an audience that same sense of wonder and enthusiasm you felt when you first took up the activity. Here are a few hobby-related subjects that might be a source of a speech topic:

- Your musical preferences and how they may have changed over time
- A favorite book or type of book or movie and why it appeals to you
- A historical period that captured your interest and why
- A sport you participated in and why it was or was not worth your time
- A highly technical hobby and how it challenged you

Choose an Academic or Career Interest

One natural choice for a speech topic in college would be your career or academic interest. Certainly this option is legitimate, but be cautious. You may not yet know enough about the field to speak about it knowledgeably and there may be others in the audience—students and instructors—who know much more about the topic than you do. That said, it certainly is possible to give a good speech on your career interest, especially if you look for an interesting angle. Here are some possibilities:

- How a favorite hobby led to a career choice
- How a work experience helped form the foundation for your career
- How a personal relationship shaped your career plans
- How a personal "calling" led to a career choice or change
- How an "accidental" course selection led to a career choice
- How a job co-op or internship led you toward, or away from, a career

Examine the Experiences of Others

When you don't have personal experience related to a topic or when you want to avoid a self-focused speech, consider conducting a kind of "oral history" project using the experiences of others. Many people can provide you with personal anecdotes, perspectives on historical events, views on current events, information about careers, and more. You may be surprised that even people you do not know you personally will be glad to assist you. Often such sources for speeches are just waiting to be asked. Here are a few possibilities, some obvious and some not:

- Parents, grandparents, and other close relatives
- Close family friends

- Professionals in the career you have chosen
- Graduates of the college or university you attend
- Nonfaculty employees of your college or university
- Older men and women with good stories to tell
- Recent or not-so-recent immigrants

Explore Possibilities in Current Events

Current events provide excellent material for informative, persuasive, and occasional speeches of all kinds. Like speeches on career topics, however, topical speeches pose some risks you need to avoid. In particular, be sensitive to topics that might offend listeners or just plain irritate them. Some people and events that might strike you as long past being controversial can still rankle many years later. There are some "hot button" topics that can still get a rise out of people, so be careful. Of course, if your purpose is to discuss a controversial topic, that's different.

Be careful to choose a topic about which you can show some depth of knowledge, especially if there is any possibility the audience will include "experts." Here are few general subjects:

- Little-known information about or an unusual slant on a current event
- Contrasting views of a current event
- An event that did not get much coverage, but which you think should have
- Your changing views of a current event over time
- An event you think will become historically important and why

Ideas for Technical Presentations

If your field of study is technical, such as electronics, engineering, computer science, or any applied science, you might want to create a presentation on a topic related to technology. Here are some ideas for technical topics:

- Trouble and how to avoid it: The importance of maintenance, testing, troubleshooting, and repairs
- Disaster analysis: What went wrong? Show cause-and-effect and explain the lessons learned.
- Controversies in technology: Explain the different points of view regarding privacy, the environment, medical research, intellectual property rights, and energy.
- Rules and regulations: Explain what engineers and manufacturers can and can't do, and why.
- An international perspective: Engineering and technology abroad; international cooperation and competition

Keep It Simple

Whatever topic you choose, be certain that (1) you can demonstrate enough knowledge to meet the expectations of your audience, (2) you have narrowed the topic so that it can be covered well in the time allotted, and (3) you believe the topic has a good chance of engaging the attention and interest of the listeners. Don't try to get too elaborate; this is one of the biggest differences between a writing and a speech. You should be able to reduce your main topic to single sentence.

3.6 KNOW YOUR TOPIC

Gather Information from Credible Sources

How you go about collecting information will depend upon how much you already know about the topic. Start your search by looking for the broadest background information that is as up-to-date as you can find. The Internet is an excellent resource for finding this background information, but remember, this is just the start.

Online Searching

If you don't know what to look for, go online and see what comes up when you use a search engine like Google, Fazzle.com, or Vivísimo.com (a clustering engine). When you find a website related to your topic, try clicking on Frequently Asked Questions (FAQ). This will give you some idea of what the website sponsors consider important issues. Following is an example of an assigned presentation on a topic with which the presenter is not too familiar:

> You have been assigned to do a short presentation on the controversy over bottled water. Because you don't know much, if anything, about bottled water or why it is controversial, you need to start by getting an overview of the topic and some idea of what the controversy is about.
>
> Cyberspace is so unpredictable and changeable that websites and search engines may or may not be there when you look for them. Here's an example of what you might find with a search for "bottled water" using Google.com and two other search engines—Fazzle.com (a very fast search engine that rivals Google in speed and scope) and Vivísimo.com (a "clustering" engine, which sorts responses into similar groups).
>
> ***Google.com***
>
> Using *bottled water* as your keywords will result in thousands of hits. Near the top of the list will be a website that says "The International Bottled Water Association (IBWA) is the authoritative source of information about all

types of bottled waters distributed in the United States. . . . " The URL for the site is *www.bottledwater.org.*

If you then go to the homepage of the International Bottled Water Association, you will find buttons for Facts and Policies, either of which might help you understand why bottled water is controversial.

Fazzle.com

A search on Fazzle.com for *bottled water* will find a site called "Bottled Water: Pure Drink or Pure Hype? While bottled water marketing conveys images of purity, inadequate regulations offer no assurance." The URL is *http://www. nrdc.org/water/drinking/nbw.asp* and the site is sponsored by the Natural Resources Defense Council. Getting this kind of result will help you understand what the debate is about.

Vivísimo.com

This is a "clustering" engine, which sorts responses into similar groups. A search for *bottled water* on Vivísimo will give you results for about 240 hits, clustered into subcategories. It is easy to refine your search by clicking on one of the clusters that better describes what you are looking for. Figure 3–2 shows what a Vivísimo search looks like.

The very first result found with Vivísimo is *www.bottledwaterweb.com*—a site called "Bottled Water Web" described as: Use the search engine or click on suppliers, bottlers, books, or industry facts. Analyzes companies and compares various water products.

By the time you have completed your basic search, you will have discovered what is controversial about bottled water. For example, some people say it contains impurities and might not be safe. Others claim it is wasteful to pour tap water into plastic bottles that are not degradable. The next

Clustered Results

- ▶ **bottled water** (239)
- ⊕ ▶ **Label, Custom** (28)
- ⊕ ▶ **Association** (14)
- ⊕ ▶ **Delivery** (21)
- ⊕ ▶ **Water Filters** (19)
- ⊕ ▶ **Tap** (17)
- ⊕ ▶ **Equipment** (15)
- ⊕ ▶ **Health** (12)
- ⊕ ▶ **Consumer** (12)
- ⊕ ▶ **Water coolers** (15)
- ⊕ ▶ **Purifiers** (14)
- ▼ More

FIGURE 3–2

Internet search results for "bottled water" with Vivísimo, a clustering search engine step in researching your topic to decide which of these issues you need to learn more about.

Online research has its limits. Here are some reasons you need to be careful with information you find on the Web:

- With billions of websites, no browser can hope to search more than a small fraction of what is online, so there is a chance that your search will omit something critical.
- Many websites are sponsored by organizations hoping to make a profit or to sway public opinion on an issue. In either case, the information is sure to be biased.
- Because the Internet is an open forum for ideas, anyone can create and post a website, regardless of their credentials or trustworthiness. Unlike articles that appear in professional journals, you can't be sure the information on a website has been peer-reviewed or evaluated by experts. Be careful, and somewhat skeptical, and you will find reliable, credible sources.
- Many organizations post press releases on their websites as a way of communicating with the public. Learn to recognize press releases (which always present one side of an issue) from serious news reporting (which strives for objectivity and impartiality).

It is a good idea to learn more about the organization or individual behind a website before you use information found there. Here are some tools for evaluating Internet sources:

- Extensions (the three letters that follow the dot in a URL) are categories of Internet domain names. Look for sites and authors that have a high degree of credibility. Table 3–1 shows what the various extensions mean for researchers:
- An excellent discussion of how to evaluate information found on the Internet, including a Web Page Evaluation Checklist, can be found at a site called *"Finding Information on the Internet: A Tutorial"* at *http://www.lib. berkeley.edu/TeachingLib/Guides/Internet/Evaluate.html*.[7]
- Once you find articles about your topic, use an Internet browser to learn more about the authors and the organizations sponsoring websites.
- Sort through the background information you find by date to get the very latest information.

Overall, it is wise to use material found on the Internet with caution.

---

[7]Joe Barker, "The Best Stuff on the Web" (accessed February 7,2005). (Berkeley, CA: University of California–Berkeley, The Teaching Library, 2002), *http://www.lib.berkeley.edu/TeachingLib/ Guides/Internet/Evaluate.html*

TABLE 3–1
Extensions Can Help You Evaluate Website Credibility

Extension	Type of site	Credibility and objectivity
.gov	Government-sponsored sites	Almost always objective, but it may be biased to reflect policy decisions made by office holders
.edu	Colleges and universities	Almost always objective, but academics have opinions and personal prejudices that may affect their research findings
.com	Commercial	Most widely used extension indicates a for-profit company sponsor
.net	Businesses that are directly involved in the infrastructure of the Internet	Internet service providers, Web-hosting companies, and so on, with an interest in making money
.org	Non-profit groups or trade associations	Promote the interests of members and advances point of view on policy issues
.biz	Small business	For-profit companies
.info	Resource websites	Provide information to the public; tend to be fairly reliable and credible

Use the Library

Your public (or college) library is one of the best places to look for credible sources of information. Professional librarians are available to help you with your search. They can direct you to credible resources you can use to support your presentation. Library resources include online catalogs and databases, access to news reports, product reviews, and information about businesses and industry leaders.

Read a Textbook

Textbooks are designed to distill a lot of information into a concise, easy-to-follow text. Like newspapers and magazines, they are written for a specific audience and age group, which means that the authors simplify concepts and use an age-appropriate vocabulary. While they are good for gathering background information, serious scholarship requires you to find original documents whenever possible. Librarians can help you locate original scientific reports.

Consult a General Reference Book

If you don't know much about a topic, it is perfectly acceptable to start learning more by using a dictionary, atlas, almanac, or encyclopædia to get basic information. However, it is not acceptable for college-level researchers to start and finish their research with a general reference book.

Refine Your Search

Once you have acquired basic knowledge, you are ready to refine your search to focus on the points you want to make in your presentations. Make notes about anything you need to know and use these to guide your research. This is how you might refine your research on the topic of bottled water:

You have determined that the controversy over bottled water has several facets. At this point, you have enough background information and can start to focus on the research you still need to do:

What you have found out so far	What you still need information about
Some criticize the lack of government oversight and worry about contaminants in the water.	Is bottled water free of contaminants and safe to drink? What government agencies are supposed to be monitoring water quality? Is bottled water different from tap water?
Others are worried that bottled water drinkers, especially children, are not getting the fluoride they need to protect their teeth from decay.	Does fluoride really work? Are there other minerals that people need in their drinking water?
Some are concerned that the huge number of plastic water bottles is filling up landfills.	How many plastic water bottles wind up in landfills? Is this a serious problem? Why aren't they being recycled?

A few people think that chemicals from plastic water bottles get into the water and may by harmful.	What are the bottles made out of? Is there any scientific evidence that chemicals from the bottles get into the water?
And the marketing of bottled water is problematic for those who feel that it encourages wasteful spending by consumers.	How does the cost of tap water compare to the cost of bottled water? What groups are being targeted by marketing campaigns and why?

Use the questions in the right-hand column to guide and to focus your research.

If you already know a lot about the topic, you should do enough research so that your knowledge of the topic is deep, current, and connected to the broader issues.

- **Deep.** You should know enough to be able to answer *why* questions and to give in-depth technical answers, if appropriate.
- **Current.** Make sure your information is the very latest so you won't be taken by surprise if you're asked about current trends.
- **Connected to broader issues.** Find out how your topic is related to other public policy issues, socioeconomic factors, environmental concerns, and so on.

Ask Experts for Information

Whether you are still in school or working in business or industry, you may find that there are people all around you who can help you learn more about your topic. If they don't have specific answers, ask them to direct you to other resources.

Don't ask for trade secrets (formulas or processes used to make a product, known only to the company that manufactures or uses them) or proprietary information. You will not get it and you will embarrass the person you ask.

Another, more objective source of information is company directories. You may want to ask a reference librarian to recommend others or to assist you in using the online versions of these databases:

- D&B Million Dollar Directory
- Moody's Industrial Manuals
- Standard and Poor's Register of Corporations, Directors, and Executives
- Thomas' Register of American Manufacturers
- Ward's Business Directory of Largest U.S. Companies

Ask manufacturers for help. Many will send you brochures, samples, posters, videos and other aids if you give them some lead time. You can use some of these for handouts or visuals.

Often your speech topics may require that you find detailed information about specific firms. Many companies now produce sophisticated Web pages about their services and products. While not without bias, they can be a good source of information.

3.7 *USING BORROWED INFORMATION*

Some speeches do not depend on published information or other borrowed material. They are delivered in your own words from notes, involve no library work, and are not reproduced as text with notes and bibliographies. This part of the chapter doesn't apply to them.

Plagiarism and Copyright Infringement

If your speech incorporates borrowed information, you should be aware of issues surrounding plagiarism and copyright infringement. *Plagiarism* is an academic dishonesty problem that involves the intentional or unintentional taking of someone else's words or ideas and passing them off as your own. In the academic world, plagiarism is a serious offense and may result in severe penalties, depending on the institution involved and the nature of the material plagiarized.

Infringement is both a civil and criminal matter involving the taking of someone else's property—in this case his or her words—and using that property without obtaining or paying for permission. While few district attorneys are willing to invest staff time prosecuting offenders, many copyright holders have their own attorneys and they are quite willing to sue violators for damages.

Failure to cite sources can do irreparable damage to your reputation and credibility. Whether you write out a speech not, you have an obligation to your audience to be clear about the degree to which you rely on borrowed information. Sometimes just a quick comment will do. Other times a more formal approach is needed.

Avoiding Plagiarism

The rule for avoiding plagiarism is this: With the exceptions of public domain and common knowledge, acknowledge all borrowed information used in your speech. Material in the public domain includes anything not covered by copyright protection, such as government publications and public records. "Common knowledge" is information generally available from several different sources in the field. Examples of common knowledge are titles of public figures, the names of the planets in our solar system, the law of gravity, how to solve a quadratic equation, or the calendar.

When you are uncertain whether or not a piece of borrowed information is common knowledge, proceed to cite the source. It is better to err on the side of excessive citation than to leave out a reference and risk the charge of plagiarism. Here are three main reasons for documenting sources thoroughly:

1. **Ethics.** You have an ethical obligation to tell the audience where your ideas stop and where those of another person begin. Otherwise you would be presenting the ideas of others as your own.
2. **Law.** You have a legal obligation to acknowledge information borrowed from a published source. In fact, if you plan to publish your speech, you should seek written permission to use borrowed information.
3. **Courtesy.** You owe the audience the courtesy of citing sources for those who may later wish to find additional information on the subject and author.

Some plagiarism occurs when unscrupulous speakers and writers intentionally steal the work of others. However, most results from mistakes during the research and drafting process, not from ill intent.

Guidelines for Using Borrowed Information

No matter what final form your speech takes, use the following guidelines when you collect information other than common knowledge.

1. **Write down and keep track of sources.** Using index cards, write a complete bibliography card for each source that you may use in the speech. A typical card might include author, title, publication information, library call number, and note to yourself at the bottom as to the potential usefulness of the source. Be precise in listing the information so that you will not have to consult the original source again when you assemble the final bibliography. *Note:* Modern electronic libraries have software that will automatically help you create a complete academic bibliography, capable of using a wide variety of citation styles. Ask your librarian about this.
2. **Take careful notes about your sources.** Most plagiarism occurs when notes are taken hastily. The important note-taking stage requires that you attend to detail and follow a rigorous procedure.

Among other things, you should (1) distinguish your own summary or paraphrase of a source from material directly quoted, (2) include exact wording of direct quotations, circling the quotation marks so that they will not be missed later, and (3) label the exact citation of the source (title, author, page, and date) so that there will be no confusion later on.

Unintentional plagiarism can occur when notes are transferred inaccurately from notecards to draft. Circle quotation marks on the draft so that it is absolutely clear what is borrowed, on the one hand, and what is either your writing or paraphrased material, on the other. Again, remember that all borrowed information must be attributed to a source unless it is common knowledge.

3. **If you use someone else's visuals, give credit to sources.** If you use visuals taken from outside sources, follow these guidelines:
   - If all your visuals come from a single source, you may cite that source at the beginning or the end of your talk by mentioning it and by writing it on one of your visuals.
   - If you use visuals from a variety of sources, print the source directly on each visual. Include enough information so audience members can find the image themselves if they so wish.

4. **If you use a direct quote, make sure you get it right and don't take a quote out of context in such a way that would distort its original meaning.** Be careful to attribute the quote to the correct source and make sure you spell the source's name correctly.

5. **If you use data, tell the audience where it came from.**

6. **Follow one of the standard citation styles.** If your presentation is based on a formal document, use one of the professional styles for citing sources, such as APA, MLA, University of Chicago, or whatever is standard in your field.

7. **When taking images or information from a website, give enough information so audience members can find the original source.** When you use information taken from the Internet, you can cite the URL as the source or find and cite the original document. The advantage of using the URL is that you may be able to add a hyperlink to your visuals that will enable you to go directly to the website during your presentation. The advantage of citing the original document is that, unlike websites, the chances are good that it will be accessible for a longer period of time.

8. **Keep track of borrowed material, so you remember where it came from.** Whether you write out the text of your speech or just produce notes beforehand, you should take care to separate borrowed information from

your own ideas, following a somewhat traditional—but still quite effective—procedure for avoiding plagiarism in speeches.

For speeches you give from notes, you can indicate your use of material that is not common knowledge by following these simple strategies:

- Tell the audience when you are shifting in and out of quoted material by using the words "quote" and "unquote."
- Give credit to sources that you paraphrase or summarize by making a general remark before, during, or after the speech—whichever seems most appropriate.
- For speeches that depend heavily on outside sources, prepare a list of "works cited" in case some member of the audience would find the list useful.

ENCOURAGEMENT

If you take your topic seriously, the audience will, too. Put energy and enthusiasm into your presentation, and your audience will be more receptive to your message.

EXERCISES

1. Observe a presentation where the speaker is an expert in his or her field. Compare the presentation with the title and printed descriptions of the talk. Do you think the speaker fulfilled his or her purpose? Explain why or why not.

TAKING IT TO THE NEXT LEVEL

It is possible to hire a consultant to help define your message. These experts can conduct surveys and run focus groups to determine the best way to get a message across to an audience. Politicians often use professional speechwriters to help them develop clear, concise, memorable ways of explaining their positions. One way to do this "on the fly" is by doing a quick survey of the audience before the presentation—ask screening questions to find out what they know, how they feel, and what they expect from the presentation. This information can be collected by a show of hands or by means of a short written instrument.

2. Record a thirty-second TV commercial that appears during one of your favorite shows. Watch it again and see if you can determine how the advertiser is trying to persuade you. At what kind of audience is the commercial aimed? Turn down the sound and look at the images on the screen; then play it again with the picture off and listen to the soundtrack. Is the ad persuasive?

3. Research a controversial topic using online and library sources. What are you able to learn about the authors of your sources? What can you conclude about the credibility of these authors?

4 Coping with Anxiety

ANXIETY: EFFECTS OF STRESS IN THE SPEECH PROCESS

At some time in their lives, most speakers feel nervous before a presentation. It's perfectly normal. Indeed, the best speakers are often those who learn to use nervousness to their advantage, rather than letting it become an obstacle. This chapter includes background information about the problem of speech anxiety as well as some concrete suggestions to help you stay calm. As you read, however, remember that you cannot eliminate all anxiety, nor should you try to do so. Some level of moderate anxiety—as opposed to "high anxiety"—keeps your speech engaging and your tone energetic.

Anxiety and Stage Fright Are Extremely Common

For many people, an instinctive "fight or flight" response kicks in when they speak. It may even engage when they first learn they will be giving a speech, sometimes building to a point at which the anxiety is far out of proportion to the importance of the speech. Fear of speaking (sometimes called *stage fright*) is so common that it is worth discussing at length.

Some studies indicate that the fear of public speaking is the most widespread of all fears, exceeding fear of heights, fear of snakes, fear of death, and more. In fact, some of the greatest performers in the world have suffered from the terrors of stage fright. Luckily, for most people, the more often they speak, the more comfortable they become in front of an audience. So, the more experience you get in public speaking, the more self-confidence you will have and the less anxiety you will feel.

Despite this anxiety about public speaking, most of us feel comfortable conversing in smaller groups, such as when talking with friends or colleagues in impromptu or scheduled sessions. We're used to situations that involve the informal exchange of ideas. Formal presentations, however, place us in a more structured, more awkward, and more stressful environment. Although we may know the audience is interested in what we have to say, the formal context triggers nervousness that is sometimes difficult to control.

To understand the different ways of coping with stage fright, you need to understand where it comes from, how it leads to stress, and some of the things you can do to cope with stress.

Anxiety Comes from Fears

The most common response people have when asked to get up and speak is fear. Some reasons people give for being afraid of public speaking include:

- Fear of being thought stupid, of making a fool of oneself, and the embarrassment, shame, and disgrace that go with that
- Fear of the audience turning on the speaker and becoming actively hostile
- Fear of being caught unprepared or of losing control of the presentation

Some of these fears are realistic, but others are not. Regardless of whether the basis for the fear is realistic or not, the feeling of fear is real to the presenter. The following three sections discuss reasons why many people experience anxiety and fear when asked to speak.

Self-esteem on the Line

When you get up to speak, there is a risk of being emotionally hurt by the audience; your self-esteem may be vulnerable to their likes and dislikes. If they like you, you may feel good about yourself. If you don't get a good response, you may feel that you are a failure. Rejection can hurt a lot. Public speaking is stressful because speakers feel that the audience is judging them and criticizing them. It is true that audience members do make judgments about speakers and that is why it is so important to make a strong, positive first impression, but knowing that doesn't make presenting any less stressful.

Jobs, Income, and Careers May Depend on a Speech

A second reason speaking is stressful is that there may be a lot on the line. The speaker's job, income, reputation, or status may be in jeopardy if the speech doesn't go well.

Promotions and raises may depend upon your coming across as the kind of person who deserves them. Your reputation as an expert may be questioned if you don't do a good job. Or a bid presentation that doesn't result in a contract may cost your company a lot of income. Too many such failures and the company might face bankruptcy. The more that is at stake, the more stressful the presentation.

A Natural Desire to Be Accepted and Liked

The need to be accepted and liked by others can add to the stress of public speaking. If you are worried about how you are perceived—worried about the audience's opinion of you—giving a presentation may be a lot more stressful than for someone who has more self-confidence. Thus, you may be stressed out by a completely normal desire to please people and to have them like you.

Fear Creates Stress

The response most people have to public speaking is a stress response. This is because they feel vulnerable and open to a wide variety of criticism. If you already have a good set of tools for coping with stress, you are well-prepared for public speaking. If you lack confidence in your abilities to speak publicly, you need to learn about different ways of coping with stress.

As already noted, your ultimate goal is to reduce your nervousness to the point where it serves your purpose—that is, where it helps you create an animated, enthusiastic presentation. Just as veteran actors use some degree of nervousness to improve their performance, speakers aiming for excellence can benefit from the same effect.

Effects of Stress

Stress (sometimes called the *fight or flight response*) often kicks in when a speaker experiences a high degree of nervousness caused by rushing adrenaline. Some of these effects of stress are obvious, but others are not:

- Dry mouth and throat
- Increased heart rate
- Short, shallow breathing
- Queasiness
- Feeling weak and unsteady—shaky or woozy
- Tightening of muscles, including those that control voice and speech
- Perspiration
- Decreased circulation in hands and feet (cold hands)
- Increased circulation in head and face (blushing)
- Tension in neck, jaw, chest, and stomach

When you consider how these changes in your body affect your ability to speak effectively, you will understand the importance of finding ways of coping with stress.

4.2 COPING WITH THE SOURCES OF STRESS

Where it was once possible for a professional to work alone, it is increasingly difficult to find any job that doesn't entail some public speaking. It is simply not realistic to expect that you will be able to find a job where you will never have to stand up and face an audience. Engineers and technical experts are now expected to be able to write and speak about what they do for a wide variety of audiences.

If you feel that making a presentation makes your self-esteem vulnerable, you can take steps to protect yourself. The most important of these steps are careful planning and getting control over your timing.

Preparation and Planning

Preparation and planning are keys to coping with the source of stress. If you are afraid of losing control of your presentation, preparation and planning will help you avoid surprises. When you take your work seriously, other people will too, so put your time and energy into planning and you will find that it is much easier to maintain professional standards and to get the kind of respect you want from your audience. You will find more information about coping with the sources of your stress in other chapters of this book:

- **Know your audience.** If you do a thorough audience analysis, you will understand who they are and what they need from you. This will make it a lot less scary to speak to them. (See Chapter 2.)
- **Define your purpose.** If you are clear about what you want to happen as a result of your speech, you are far more likely to come across as confident and competent. (See Chapter 3.)

- **Practice.** The more you know what is supposed to happen during your presentation, the less likely it is that things will go wrong. (See Chapter 5.)

Get Control of Your Timing

Nothing will alienate your audience more quickly than a perception that you are wasting their time. Respect your audience by starting on time and finishing on time (or early).

How long is too long? Everyone speaks at a different pace, so it is hard to give a rule for how long a practice run should go in order to fit into the time allotted. If you practice with a timer, you will have a better sense of how long your talk will go, but the length of your practice runs may not correspond to the length of the actual presentation. Some beginning presenters run short and others run long when practicing their presentations. The point is that you should practice enough to know which problem you have. You will get better at timing the longer you work at preparing and presenting.

Following are some examples of how some speakers lose control of their timing, along with some solutions:

- **Late start.** Some people seem to have a terrible time getting going, whether because they are unfamiliar with projection or sound equipment, have trouble finding their visuals, are generally lax about sticking to schedules, or have some other problem. Even if no one makes a fuss, if all the other speakers do it, and if it seems acceptable to start late, professionals who respect their audiences will make every effort to get back on time. Audiences don't always treat all presenters equally—the same group that will sit quietly for twenty minutes while a Nobel laureate fumbles with his notes will rip you to shreds in their evaluations if you don't start on time.

To guard against a late start, get there early and have everything set up and ready to go when the clock strikes your announced starting time.

- **Rambling introduction.** Sometimes you run into a speaker who feels a need to connect his or her topic with recent world events (where "recent" includes the American Civil War) or who decides to establish credibility with the audience by reciting each and every article, grant, and research project to which his or her name was ever attached. The audience decides quickly that the presenter is in love with the sound of his or her own voice. Even the most polite people will shift uneasily in their seats until the ego-obsessed speaker gets to the main topic, if ever.

  To keep from rambling, get through your introduction quickly and move into the body of your talk.

- **Startled by the unexpected.** Speakers are sometimes caught unaware by a question or other interruption. Some are better than others at getting back on track. A few will hem and haw and try to work their way out of an awkward situation by jumping from point to point with "and another thing is that . . . ," desperately seeking an answer. Eventually they stray so far from the path that they completely lose sight of what they came to talk about.

  Do enough research so that you are unlikely to be caught off guard and, if someone tempts you from your path, refuse to follow. Instead, note that this is an interesting point and that you hope to do more research on it in the future. Then get back to your topic.

- **Presentation is too short; there is time left over.** Every so often a presenter discovers that he or she has more time to talk than expected. It is always a good idea to have some extra material at hand in case you are given more time. When you prepared your talk, you probably had to shorten some parts to fit into the time allotted. This is your chance to use that material. Extra time gives you an opportunity to go into your topic in greater depth. Here are some ideas for how to use your extra time:
  1. Make up a handout with extra material just in case you get more time. Use your time to discuss it.
  2. Think up some intriguing questions about your topic and discuss them.
  3. Describe some of the leading people in your field and the kind of work they are doing.

4. Speculate on the future of your topic; what might it be like in twenty years?

5. If you have a computer/projector with Internet access, show your audience some relevant websites.

6. For younger audiences, describe how you got interested in your topic and the educational preparation required for someone to get involved in your field.

■ **Presentation is too long; you run out of time.** You have been asked to share the podium with two other speakers for an hour, and each of you will have twenty minutes to speak. The first speaker rambles on for half an hour and then answers questions for five minutes. The second speaker uses up the allotted twenty minutes. You have five minutes to do a twenty-minute talk: What do you do?

Resist the temptation to strangle the first speaker. Instead, give a brief overview of your main points and explain that more information can be found on the handout you brought with you. If you can, offer to stay around to answer questions for as long as possible.

■ **Unexpected sidetracks.** Every so often a speaker will discover too late that something he or she thought was common knowledge among audience members isn't well known at all. A speaker who is attentive to audience responses will soon realize that they are not following along. It might be the result of an allusion, the use of jargon, or a reference to recent scientific discoveries, historic events, or key individuals.

Even professional speakers who are well prepared and who think they have thought through every aspect of a topic in advance may still misjudge an audience's breadth of knowledge.

1. Find out how big the problem is. Ask a screening question to determine how many audience members are lost.

2. Decide how extensive an explanation is required. If the term is key to your talk, you will need to spend more time explaining it. If it is merely incidental, give a one-sentence definition and get back on track.

Example: In a talk about migraine treatment and prevention, the speaker detects some puzzled looks among audience members. "How many of you are familiar with serotonin and what it does?" If understanding serotonin is a key to following the rest of the presentation, the presenter will have to go into some detail about how it acts as a neurotransmitter and modifies the effects of other neurotransmitters. If it is merely incidental to the talk, then "it's an important brain chemical believed to play a role in causing migraines."

When you organize the timing of your talk, keep the following guidelines in mind:

■ Don't try to make too many points in a short period of time. That's how a lot of presenters get into time trouble.

THINGS THAT DON'T REDUCE STRESS

Because there is a lot of bad advice floating around about stress reduction, here are some things that *will not* reduce anxiety and stress:

- Refined sugar, which will give you a short burst of energy, followed by a period of sluggishness as your body adjusts
- Television commercials, which are designed to increase viewers' level of arousal, in other words, to create stress
- Driving: When you are preoccupied with something stressful, you are not paying full attention to the road, which is dangerous.
- Alcohol
- Drugs, including narcotics, "uppers," and marijuana
- Nicotine

Here are some things that *may affect* your feelings of anxiety:

- Caffeine: Too much makes some people nervous and giddy.
- Carbohydrates: Too much before presenting can make you drowsy and slow you down.
- Aspartame: This low-calorie sweetener (often in light-blue packets) is used in a variety of foods and drinks and as a tabletop sweetener. It is about 200 times sweeter than sugar and is commonly known by names such as NutraSweet, Equal, Spoonful or Equal-Measure. Many people report that this artificial sweetener makes them have to urinate more frequently, which may distract you and disrupt your presentation.[1]

[1]Claudyne Wilder, "Presentation Points," *www.wilderpresentations.com* (accessed June 7, 2004).

- Practice with your visuals and time yourself to get a better idea of how long your talk will last. A "dry run" without visuals will only give you the vaguest idea of how long your presentation will go.
- If you are presenting as part of a team, make every effort to practice together before your presentation. If you can't practice with your team, make sure everyone knows exactly how much time he or she has to speak and agrees to stick to that schedule.
- Don't rush. Work on smooth pacing.
- Don't test your audience's endurance. Build in one ten-minute break for every seventy-five minutes.

- Don't start the question-and-answer session when your talk is scheduled to end. If you plan to have a question-and-answer session, allow time for it at the end of your talk.
- Distribute handouts with a "barebones" outline of your talk so audience members can follow your progress and see where you are.

4.3   COPING WITH THE EFFECTS OF STRESS

Use the energy and alertness that result from stress as tools to help you give a better presentation. As the cliché goes, don't try to eliminate all the "butterflies" before a presentation; just get them to fly in formation. A certain level of nervousness will remain even after you've applied the guidelines in this chapter. Then you can go about the business of getting it to work for you, not against you. Note that some cases of stage fright may be so severe that they require therapy or prescription medication (beta-adrenergic blocking agents).

The following sections discuss some guidelines for coping with the effects of stress that will help most people get over their anxiety.

Prepare Your Presentation Well

The most obvious suggestion for eliminating nervousness is also the most crucial one. Work hard to prepare your speech so that your command of the material will help to conquer any queasiness you feel. Besides making you feel more confident, knowledge of the material helps listeners overlook any nervousness you think may be evident.

Although this guideline is one over which speakers have the most control, many disregard it. Some believe they know the material so well that they don't need to practice the speech. After all, haven't they been working in this field all of their college years or throughout their careers? This attitude can result in a long, rambling, disorganized speech.

A second group of speakers are so nervous about the speech that they reach a point of resignation. They believe they are beyond help and that nothing but a miracle can save them at this point, so they develop a sense of doom that aggravates any anxiety that may develop.

A third kind of speaker knows the material and has a good sense of what he or she wants to say but is afraid that too many practices will make him or her sound stale. While it is true that reciting the same speech over and over can deplete its freshness, it is also true that, for someone who is not an experienced presenter, there is simply no such thing as overpreparing. Professional actors can speak the same lines night after night and never have it sound old, but, unless you are a theatrical performer, don't use this as an excuse for not practicing.

GETTING FIRED UP

Up until thirty years ago or so, athletic coaches believed that the best way to prepare for competition was to get "fired up" beforehand. They would give stirring speeches, followed by chants of "Go, go, go" or "Win, win, win" with the whole team screaming and jumping up and down. Getting all stressed out was considered a good thing, and the more you shouted and jumped around, the more you were considered ready to compete.

Athletes today would never prepare for a game or a meet that way. The emphasis nowadays is on stretching, warming up, and mental preparation, especially visualizing success. Controlling the effects of stress is a big part of what good coaches and trainers focus on. This change in attitude is evidenced by frequent mention of the "mental game" by sportswriters and commentators.

The same theory applies to public speaking—avoid getting yourself all stressed out and work on warming up and visualization.

Resist the temptation to "wing it" in a speech. Instead, prepare and practice until you know your material inside and out. Your high level of knowledge will add to a level of confidence to combat nervousness.

Prepare Yourself Physically

You may know of athletes who follow a certain routine before a performance, such as eating a particular meal they think best prepares them for competition or warming up with a particular sequence of exercises. Some successful speakers view the physical preparation before a speech in the same way.

The keys to physical preparation are to use fluids, diet, rest, and sleep, and exercise to help you reach a high level of alertness without increasing the effects of stress on your body and mind.

Hydrate

Drink several glasses of water in the hours before you speak. When you properly hydrate your body, your vocal cords are most likely to operate without strain during a presentation. For obvious reasons, don't take this suggestion to the extreme by drinking excessively before speaking.

Watch Your Diet

Avoid caffeine or alcohol for several hours before you speak. Both of these—even in small amounts—can alter your responses, either by slowing down your reactions, in the case of alcohol, or speeding them up, in the case of caffeine.

Just as important, they may add to your anxiety by causing you to question whether, in fact, they might be influencing you to some degree. Again, avoid any substance, activity, or thought that may reduce self-confidence.

In addition, eat a light, well-balanced meal within a few hours of speaking. Most people prefer a balanced meal an hour or two before a speech to keep the body functioning smoothly. However, don't overdo it by eating heavily—particularly if the meal comes right before your performance as a featured speaker.

Exercise

Exercise normally the day of the presentation. A good walk, for example, may help to invigorate you and use up the adrenaline that leads to nervousness before a presentation. Or you may prefer more vigorous exercise such as swimming, hiking, or jogging. However, don't wear yourself out by exercising in what would be an unusually excessive fashion for you.

Get Enough Rest and Sleep Beforehand

Nothing will mess up the most carefully planned presentation more than for the speaker to be sleep-deprived. Make it a rule to always get a full night's sleep for a night or two before you have to speak. This is especially important if you have traveled across time zones and may be suffering from jet lag.

Relaxation Exercises

Perform deep-breathing exercises before you speak. When a person is under stress, breathing becomes shallow and rapid—two conditions that disrupt delivery and make it hard to speak in a loud, clear voice. Breathing deeply and controlling your breathing will reduce the effects of the butterflies in your stomach.

Posture is important in deep breathing, so stand up as straight as you can. Become conscious of your breathing and then work at inhaling and exhaling slowly. Always breathe from your diaphragm, the muscle below the chest that, when lowered, allows the lungs to increase the amount of air they take in. If you hold your hand over the middle of your abdomen, you should be able to feel your diaphragm moving up and down, filling your lungs. Your goal is to gain control over the intake and outflow of air so that you train your body to slow down to a pace you can control. When you do this, you will know you aren't taking in too little air and then pushing out whatever you've got left too quickly.[1]

---

[1]Kirby Tepper, "Breathing," *http://www.powerpublicspeaking.com/breathing_exercises.htm* (accessed July 21, 2004).

Breathing exercises can be especially valuable if you do them on a regular basis. They can help you:

- Project your voice
- Keep from straining your voice
- Increase your stamina

Exerting some control over an activity that is considered involuntary takes your mind off the speech. It brings your body into a state of rest, not unlike that of Zen monks who meditate while sitting cross-legged on straw mats. We can learn from the ancient traditions.

Try Visualization—Picture Yourself Giving a Great Presentation

This guideline is a variation of an athlete's efforts to win the "mental game" before competing and is analogous to the motivational technique of "visualizing" success. The point is to flood your psyche with images of giving a successful presentation before a group of listeners who are being informed, persuaded, or entertained by your words. Whatever other value this technique may have, it certainly does take your mind off anxiety by keeping you preoccupied with the effort to visualize.

Visualization (sometimes called *imaging*) works because you program success into your thinking before you speak. As part of visualization, imagine yourself giving a successful speech, by going through the following sequence:

1. Arrive at the room with prepared materials in hand.
2. Check the visuals to see that equipment is functioning properly.
3. Exchange pleasantries with other presenters or members of the audience.
4. Rest comfortably in your chair while you are introduced.
5. Stand up and stay silent for a few seconds while you prepare to start.
6. Begin your speech with a smile and an attention-getting introduction.
7. Maintain eye contact and an appropriate pace throughout the speech.
8. Conclude with an effective wrap-up.
9. Respond with clarity and thoroughness to all questions.
10. Re-take your seat and know that you have "delivered the goods."

The experience of many who have used visualization suggests that it helps control negative thoughts that pass through the minds of even the best speakers. If nothing else, it occupies your mind with a sequential process at a time that fear and feelings of anxiety might dominate your mind.

Arrange the Room the Way You Want

A speech presents you with one of the few communication contexts in which your comfort is more important than anyone else's. Part of your ritual to prepare for a speech and, thus, reduce your anxiety, should be to assert control over the physical environment in which you speak. Here are some actions you can take:

- Set up chairs in the pattern you prefer.
- Position the lectern to your taste.
- Make certain the lighting is adequate, especially near the lectern.
- Check the position of the equipment so that it's ready for immediate use.
- Ask that the temperature in the room be adjusted for your comfort. If you must choose between warm and cool, pick cool. Although you certainly don't want your audience to be shivering in their chairs, a cooler room keeps an audience more attentive than a warmer one, which induces drowsiness.

Be Ready for Emergencies

Having back-ups in case Murphy's Law applies and things go wrong may reduce your anxiety about speaking. The "be prepared" school of public speaking suggests that it is worth considering anything you can do that reduces your anxiety and increases preparedness.

- For speakers who are a bit nervous, having water available is important because a dry throat often accompanies nervousness before and during a presentation. To help them stay hydrated and to combat the effects of a dry mouth, experienced speakers often keep a glass of water at hand when they speak. Speakers usually take a drink of water at strategic points, such as while a question is being asked or while the audience is applauding.
- Never assume that your visuals will work the way they are supposed to. Always make an extra set in a different medium. For example, if you are planning on running your visuals from a laptop computer, protect yourself by copying them onto a CD, printing them out on paper or transparencies, or attaching them to an e-mail that you send to yourself.
- Have a supply of extra handouts to use in case the equipment required to project your visuals malfunctions.
- Bring along extra marker pens for the overhead projector, smart board, or flipchart—just in case you need them.

Talk with Audience Members Before Your Speech

Casual conversation with the audience often helps reduce the distance you feel between you and the listeners. This distance often presents a psychological barrier to speakers who envision their audience as potential critics (or even "the enemy"). Almost all listeners empathize with speakers and want them to do well, and you become acquainted with this goodwill when you talk casually with them before the speech. In addition, you get a chance to exercise your voice and calm your nerves. It then won't seem as much of an abrupt transition when you begin your formal speech.

Remember That You Are the Expert

When you deliver a speech in class or at work, remember that you are providing information that your listeners do not have. Reminding yourself that you are the one with expertise on the subject should help to boost your confidence. Your listeners want to hear what you have to say. Tell yourself, "I'm the expert here!"

Don't Admit Nervousness to the Audience

No matter how anxious you feel, resist the temptation to admit it to others. First, you don't want listeners to feel sorry for you—that won't lead them to a positive view of your speech. Second, you may be surprised to learn that your nervousness is almost never apparent to the audience. Studies have shown that people are good at recognizing emotions like anger, joy, and distress, but they are very poor judges when it comes to recognizing fear. (In fact, one of the most difficult things for an actor to portray is fear.) Although your heart is pounding, your throat feels dry, and your legs are a bit wobbly, few, if any, listeners will observe these symptoms. If you don't mention it, they will never know, so why draw attention to a problem that may be unnoticed?

Slow Down

Out of nervousness—and from a desire to get out of the spotlight as quickly as possible—some presenters speak faster during the actual speech than they do during practice. Because this increased pace often appears obvious to your audience, it can reduce your effectiveness as a speaker. Here are some strategies for preventing it:

- Practice with audiotape or videotape so that you are very familiar with how your voice will sound in a formal presentation.
- Ask someone for help if you need an audience for your practice runs. An important presentation should not be a one-person job.
- Constantly remind yourself to slow down during the speech and to use pauses for rhetorical effect.
- Maintain strong eye contact with your audience—some speakers begin to talk too fast when they reduce eye contact with the audience and focus on notes.

- When practicing, write down reminders, such as "Slow down here!" or "Whoa!", on the copy of your speech. However, do *not* put these reminders on the note cards or script you use during your actual speech or you might read them aloud by mistake.
- Time yourself in practice, and then note in the margin of the speech text about where you should be at certain intervals (of course, any looking toward your watch or a clock during the speech itself should be subtle).

Seek Opportunities to Grow as a Presenter

The more public speaking you do, the more comfortable you will become. You do not need to go it alone in your effort to overcome anxiety in speaking. Find friendly audiences where you can work on improving your speaking techniques. Here are a few ways you can get assistance:

Join an Organization That Promotes Speaking

- Toastmasters International (*www.toastmasters.org*) has chapters all around the world and encourages new ones to form where needed. These clubs meet regularly to give members the opportunity to develop and practice their speaking skills in a supportive environment in which everyone wants to improve. Consult your phone book or check the Internet to contact local Toastmaster chapters.
- Society for Technical Communication (STC) (*www.stc.org*) is an individual membership organization dedicated to advancing the arts and sciences of technical communication. It is the largest organization of its type in the world, with 25,000 members, including technical writers and editors, content developers, documentation specialists, technical illustrators, instructional designers, academics, information architects, usability and human factors professionals, visual designers, Web designers and developers, and translators—anyone whose work makes technical information available to those who need it. STC has chapters all over the world. At monthly chapter meetings, STC members and invited guest speakers give educational and informative presentations related to communication skills.

Start Your Own Informal Speaking Organization

You don't need the umbrella of an international organization to start your own group. Once you have collected a few colleagues who want to improve their speaking skills, set some rules and agree to meet at a regular time without fail—such as every week or two for lunch in the company training room. One

TAKING IT TO THE NEXT LEVEL

The more control presenters have over the environment, the less anxiety they feel. Professional speakers make sure that everything about the presentation that can be controlled *is* controlled. Sometimes they have an advance person or team set up things for them ahead of time. Having an assistant (volunteer or paid) to deal with details can free a presenter to focus on the presentation.

or two members can give different types of speeches, followed by constructive discussion by all those present. There is no better way to gain confidence than by speaking more often.

Get Medical Help

If speaking anxiety presents serious physical symptoms that concern you, such as a rapid heartbeat, you might want to visit your family doctor for advice on possible medications for such symptoms.

In summary, a moderate case of nerves before a speech helps you develop a level of enthusiasm that is welcomed by listeners. However, when you have anxiety well beyond what you think is useful to your purpose, employ the techniques in this chapter to get your butterflies to "fly in formation."

ENCOURAGEMENT

You may feel frightened, but the audience won't know it if you don't tell them. Studies show most people are not good at recognizing when another person is afraid. It is very unlikely that your audience will know how anxious you are, so don't worry about it! Do not point out problems to your audience or disparage your visuals. If something goes wrong, just keep going—most audience members won't catch it or realize that you made a mistake.

EXERCISES

1. Try some of the techniques for reducing anxiety and stress described in this chapter. See if any of these help lower your blood pressure, alter your breathing, and change your voice tone. Think about any other tools you

have available for coping with stress. Write up your experience and discuss it with the rest of the class.

2. Visit a classroom, lecture hall, or auditorium where you might be giving a presentation some day. Look around carefully. Measure the space. Try out the seats, lighting, sightlines, acoustics, and A/V equipment. What do you think it would be like to do a presentation there? Write up your visit, describing the venue and what you think is good and bad about it.

5 Organizing Your Presentation

Every method of rational argumentation requires careful attention to organization in order to persuade an audience of the speaker's credibility and competence. This chapter describes how to develop an outline for a talk designed to have an introduction, a body, and a conclusion. The last part of the chapter deals with how to create a set of presenter's notes and how to practice before you have to give your speech.

5.1   OUTLINES

The process of developing a presentation starts with a purpose statement and an audience analysis. From there, you can set down your ideas for main points and organize them into an outline. This chapter explains how to create an outline following the Introduction-Body-Conclusion format, how to turn that outline into a usable set of presenter's notes[1], and how to use your outline to manage your presentation time efficiently.

Work from a Strong Outline

It is impossible to put too much time and effort into writing and polishing your outline. The more energy you devote to the organization of your presentation, the greater your chances of success. To put it another way, you can survive if your visuals don't show up, if your handouts are messy, and if they misspell your name in the program. But if your outline is weak, you're a goner.

Your outline will help you come up with a clear, logical development of your topic. Just don't wait until the last minute to start working on your outline; give yourself time to research the parts you are not sure of and to revise as your ideas become clearer.

---

[1]Presenter's notes are those notes you have in front of you when you give your talk. They may be anything from a few key words on a sheet of paper to a color-coded set of index cards. There is no one right way to make presenter's notes—whatever works for you is right.

Writing a Hierarchical Outline

The easiest way for most people to organize thoughts and to organize a set period of time is to write a hierarchical outline, one in which items are arranged in rank order. Outlines help you understand the "Big Picture"— how your talk works as a whole.

Outlining is a form of linear thinking that enables you to determine how the whole presentation flows from beginning to end. Think in terms of creating a hierarchy of ideas; some ideas become main points, others will support and expand on the main points. Not everyone is comfortable with a linear approach. Sometimes other techniques, such as mind-mapping[2] or clustering, are more effective for these people than traditional outlining.

You may feel better writing out your outline by hand on paper, but using a computer to do this has advantages over writing it out because it is much easier to revise and rearrange the different parts. You may start either with a word-processing program (such as Microsoft Word) or with presentation software (such as Corel Draw, Microsoft PowerPoint, or Apple Keynote).[3]

Outlining Methods

The process you follow will depend on the method you choose for writing:

Starting with Handwritten Notes

There are a number of advantages to making handwritten notes:

- You can do your writing anywhere.
- You can write in pencil and erase if you need to.
- Your writing will be quick and materials are cheap.

There are, however, a couple of disadvantages to going with this method:

- It is efficient as long as you have clear handwriting—if you can't read your notes, it's a waste.
- Revising sometimes mean rewriting whole sentences and paragraphs. You will probably have to rewrite the whole thing if you want a clear, easy-to-follow set of presenter's notes.

---

[2]Mind mapping is the creation of Michael Gelb, author of *Present Yourself.* Mind mapping is not for everyone, but if you are interested in trying this method, go to the website at *http://www.mindjet.com/index.shtml.* There you will find examples of mind maps and you can download free samples of software that allow you to try mind mapping.

[3]Microsoft PowerPoint dominates the world of presentation software, but it is not the only product available for illustrating presentations. If you search online for "presentation software," you will discover dozens of products that promise to give you the same tools you will find in Power-Point. The Resources section of this book presents more information about alternatives to PowerPoint.

OUTLINING WITH POWERPOINT

Microsoft PowerPoint has a dominant share of the presentation software market, so it is worth mentioning how it can be used to create a presentation outline.[4]

When you work with Microsoft PowerPoint, you have several choices of "views" for creating a presentation. Each view shows different parts of the presentation. Use **Outline View** (Figure 5–1) to type the text you want on your visuals and you will have an easy way to rearrange bullet points, paragraphs, and slides. (**Outline View** is the same as Outline pane, except that when you use **Outline View** you can't see the visuals or presenter's **Notes** as you work on the text of the outline.)

You are supposed to be able to use the **Outline View** to organize and develop the content of your presentation. However, *everything* you type into your **Outline** will appear on your visuals, whether you want the audience to see it or not. If you want to make notes to yourself about when to pause, to catch your breath, to ask if there are questions, or to give a verbal example, you can't put this into your PowerPoint **Outline** or the audience will see it. Instead, type or paste these notes into the **Notes pane.** The drawback with the **Notes pane** is that when you use your computer to project slides (using the Slide Show feature of PowerPoint), you can't see your notes. To see your notes and run the presentation visuals at the same time, you must remember to print out the Notes before you start your presentation.

You can get complete directions for using the **Outline pane** or view if you work with PowerPoint by clicking on the Help menu and typing in outline.

FIGURE 5–1
Outline View button

Click on this icon, found in the lower left corner of the screen, to see Outline View in Microsoft PowerPoint.

---

[4]There are different versions of PowerPoint, so instructions that work with one version may not work with another. See Chapter 6 for more information on using PowerPoint.

Starting with a Word-processed Document

Word-processing software offers a number of advantages as a note-writing method:

- It is very easy to revise and rearrange sections as needed.
- Most word-processing programs include an "outline" format for documents that makes it extremely easy to move things around and to get an overview of the whole.
- You won't be distracted by visuals as you write.
- This is the best method if you plan to publish your speech.
- If you are one of those people who need to write out your entire speech first, working with presentation software will be difficult for you to use and you will do a better job if you start with a word-processed document.

The disadvantage to using word-processing software to write your notes is that if you want to make visuals, you will, at some point, have to transfer your outline to presentation software. Transferring from one to the other is usually not a problem, but Murphy's Law says that the one time you must depend on the software to work properly, it will let you down.

Starting with Presentation Software

The advantages of writing an outline using presentation software is that your outline instantly becomes a set of visuals. However, because the software is so much fun and so easy to use, it is very easy to be distracted by its powerful graphics capabilities. Thus, beginners can fall into the trap of devoting all their preparation time to creating visuals instead of focusing their attention on the logic and flow of the presentation.

Guidelines for Writing and Using Outlines

Here are some guidelines for writing and using outlines.

Place Your Purpose Statement at the Top

Your purpose statement should be a single sentence, although it may be a long one. (If you can't express your purpose in one sentence, go back to Chapter 3.) Write down what you hope will happen when your talk is over. What do you want the audience to think, feel, or do when you are done? Figure 5–2 shows this first step with an outline for a presentation on hybrid-electric cars. Remember to revise your purpose statement if you later decide to change the focus of your talk.

Describe Your Main Points

The longer you have to talk, the more information you can impart; however, there are limits to how much an audience can absorb at a single sitting, so don't try to cram in too much. (You may have experienced lectures in which

> **Purpose:** To pursuade audience to consider buying a hybrid-electric vehicle for their next car
>
> **Introduction**
>
> **Body**
>
> **Conclusion**

FIGURE 5–2
Outline with purpose statement

a professor tried to cover too many topics in a single class period. One sure sign that a professor is trying to cover too many points is when students taking notes switch to "automatic"—writing as fast as they can rather than thinking about, and absorbing, the ideas presented.)

If you can limit yourself to two or three main points in the body of your presentation (regardless of how much time you have been given to speak), you will do a much better job of presenting than if you try to cover too much. It is acceptable to put additional points in a handout if you think your audience needs to know about them. Figure 5–3 shows what an outline looks like at this stage.

Each of your main points will become a part of the body of your presentation and a first-level heading in your outline. (The next section explains how to work with an Introduction-Body-Conclusion format for your talk.)

Record Your Ideas

At this point, the most important thing is to *brainstorm*—setting down as much as you can as quickly as you can. Don't spend a lot of time perfecting your ideas; that will come later. Make note of items that will require more research and don't worry about things you don't know yet.

Try to fit each idea into one of the main points of your talk. As you add ideas to your outline, it will start to look more like the example shown in Figure 5–4. If an idea doesn't seem to fit anywhere:

- It might be an interesting, but nonessential, part of your talk, so consider using it in your introduction or your conclusion.

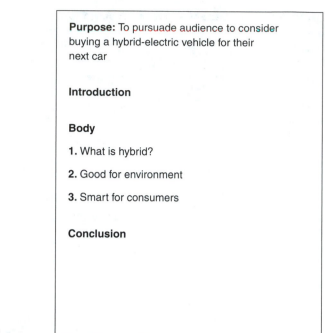

Purpose: To pursuade audience to consider buying a hybrid-electric vehicle for their next car

Introduction

Body

1. What is hybrid?

2. Good for environment

3. Smart for consumers

Conclusion

FIGURE 5–3
Outline with main points

- It might be important to adding depth to your presentation, such as explaining other points of view or the history behind your topic, so consider putting it into a handout that audience members can pick up at the end of your talk.
- It might not belong in your talk, so leave it out.

Plan Where to Use Graphics Early On
Use visuals to help your audience understand concepts, not for the sake of using visuals. Make notes in your outline about what kind of visuals you want to use and where you want to use them, as shown in Figure 5–5.

- Data graphics help explain numerical information
- Pictures help audience see what you are describing
- Keep text to a minimum

Plan Transitions from Section to Section
Consider ways of transitioning between sections so that the presentation flows smoothly. Make transitions obvious, so the audience understands that you are moving to a new part of the presentation.

Build in Pauses
Plan on giving yourself time to change visuals, sip water, take questions, and catch your breath. Pauses also give the audience a chance to absorb what you have just said.

Purpose: To pursuade audience to consider buying a hybrid-electric vehicle for their next car

Introduction

Body

1. What is hybrid?

    a. Internal combustion engine plus electric motor

    b. Sometimes runs on gas, sometimes on motor (find out when)

    c. Find out names of models on market now

2. Good for environment

    a. Lower emissions (find out how much and what types)

    b. Smog problem in cities could be reduced (find predicted impact of hybrids)

    c. Find opinions of environmental groups

3. Smart for consumers

    a. Gas prices go up and down (find out history and predictions)

    b. Lower expense in long run

    c. Trendy image

Conclusion

FIGURE 5–4
Outline with subpoints filled in

Revise: Go from Draft to Final Outline

Consider the first version of your outline a "work-in-progress" that can be changed and revised as needed. The more time you can put into perfecting your outline, the better your presentation will be.

Purpose: To pursuade audience to consider buying a hybrid-electric vehicle for their next car

Introduction

Body

1. What is hybrid?

    a. Internal combustion engine plus electric motor

    🖥 Sketch of hybrid

    b. Sometimes runs on gas, sometimes on motor (find out when)

    c. Find out names of models on market now

    🖥 Pictures of hybrids

2. Good for environment

    a. Lower emissions (find out how much and what types)

    🖥 Graph showing comparisons

    b. Smog problem in cities could be reduced (find predicted impact of hybrids)

    c. Find opinions of environmental groups

3. Smart for consumers

    a. Gas prices go up and down (find out history and predictions)

    🖥 Graphic showing trends

    b. Lower expense in long run

    c. Trendy image

    🖥 Pictures of future hybrids

Conclusion

FIGURE 5–5
Outline with ideas for visuals filled in

Outlines Help You Manage Your Time Efficiently

When you outline your presentation, consider the Introduction-Body-Conclusion format (described later in this chapter) as you plan how you will use your time. The largest block of time is for the body of your talk, when you explain your main idea(s). Allow time at the beginning for your introduction, including an audience "hook" and a forecast of where you are going with your presentation. Allow time at the end for your conclusion and call for action. Leave some time after the end of your talk for questions and answers, but not a lot. If your presentation is going to be more interactive than a straight lecture, leave time for discussion and activities. If you give out your name and number, people who really want more interaction will find a way to meet or contact you. If you can get all of this done and finish a few minutes early, your audience will love you.

Uses for Your Outline

Once your outline is complete, you can use it to create a script, a published version of your speech, presenter's notes from which to work, and a handout for your audience.

- **Create a script for your speech.** Some presenters are very anxious unless they know exactly what they are going to say, so they need to start with a complete script. If you feel you need to choose every word in advance, go ahead and write a script for your talk. When you feel comfortable with what you have written, save the document on your word processor and make a copy. You can then turn the script back into an outline or a set of presenter's notes, which will help your presentation sound fresher and more extemporaneous.
- **Publish your speech.** Use your word processor to turn your outline into a document you can publish. Follow one of the standard styles for citing sources of borrowed material.
- **Make presenter's notes.** Convert the outline into a set of presenter's notes that you can use when you do your presentation. Making and using presenter's notes is described later in this chapter.
- **Create a handout.** One of the best uses for an outline is to create a handout that will enable your audience to follow your talk and take notes. Do this by reducing what you have written for yourself to a "bare bones" outline for your audience. Give them just the most important points and leave out details. (If you hand them your entire outline, they may see no reason to sit and listen to your talk.) Leave a lot of white space between points on the handout and they will have room to make notes.

PRESENTING PROFESSIONAL OPINIONS

When you do a technical presentation, you are the expert and your audience wants you to tell them:

- What you think is causing the problem, based upon your technical expertise
- What can be done about it, based on your experience, training, and research—they want you to recommend a particular solution

Keep in mind the audience's level of sophistication when offering your professional opinion. If you lapse into jargon or come across as arrogant or pompous, you will lose your credibility as an expert.

Here are two cases in which you might be speaking as a technical professional. In the first, your purpose might be to show a link between cause and effect. In the second, you would be explaining a problem-solving process.

Linking Cause to Effect

A technical professional gathers evidence, weighs it, and draws conclusions about cause and effect. *Cause-and-effect* is used to show connections, as in offering a solution to a problem. This is how experts offer a *professional opinion.* Your job is to explain what happened and to offer a solution: If your explanation of the cause isn't plausible, they won't give you any credibility.

As an expert, you are paid to evaluate evidence and reach a conclusion. However, your explanation must be based on solid research and you must present strong evidence to support your conclusion:

- Evidence must be *adequate*—your test sample must be big enough to be valid.

5.2   USING THE I-B-C FORMAT

In most cases, it is appropriate to think in terms of a technical presentation having three parts: a beginning, middle, and an end.

Introduction-Body-Conclusion: The I-B-C Format

One main feature separates written documents from oral presentations. While reading a written report, readers control the pace at which they read. They can navigate within the document at will—rereading portions if necessary. While listening to a speech, however, listeners are powerless to control

- Evidence must be *representative*—if your test sample is no good, the evidence is useless.
- Evidence must be *plausible*—the link between cause and effect must be logical and reasonable.

You must show that existence of one thing requires the existence of the other. If you find more than one possible cause, you must explain how you, as a professional, weighed the various factors to reach a conclusion.

Explaining a Problem-solving Process

Sometimes it is necessary for technical presenters to show an audience how they think a problem should be solved by describing the process they went through.

Here is a model for organizing a presentation that shows an audience a problem-solving process. This model may remind you of the methodology used by physicians (symptoms, diagnosis, prescription, prognosis) and this process is the same one used to develop an engineering proposal.

- **What's the problem?** Give a description of the symptoms and show why the problem is significant and worth solving.
- **What's causing the problem?** Explain your problem statement and show a cause-and-effect relationship.
- **What solutions are available?** Discuss what your research shows—what solutions have been tried and which have been shown to be effective.
- **How did you select the best solution?** Describe how you set criteria and compared options and demonstrate how you picked the best one.

the pace. They have no chance to "rewind the tape" if information is missed. This basic difference between reading documents and hearing speeches suggests that you must organize speeches carefully. The audience should be able to absorb your message the first time through, because (unless you have taped it) there may be no other opportunity.

To give listeners a structure for capturing information, use a three-part structure called the *I-B-C format* (Figure 5–6), which has the following sections:

- Short introductory section called the *introduction*
- Long discussion section called the *body*
- Short wrap-up section called the *conclusion*

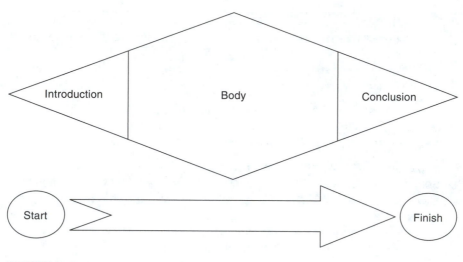

FIGURE 5–6
I-B-C format for presentations

Background on the I-B-C Format

Well over 2,000 years ago, Greek philosopher Aristotle (384–322 B.C.E.) wrote that the structure of a story forms its "basic principle, the heart and soul." He noted that plot structure must include three separate but integrated parts: the "beginning, middle, and end." It also must describe "an action which is complete" and "should neither begin nor end at any chance point."[5] Although he was speaking about the plot of a tragedy, Aristotle's guidelines apply just as well to oral presentations.

Aristotle's idea about a clear "beginning, middle, and end" serves to support the simple I-B-C format, which rests on five thoughts about the preferences of listeners:

1. Most people prefer a clear, linear, mental outline into which to place information provided in a speech.
2. People's experience with traditional print and electronic media—books, newspapers, film, and television—has shown them how communications work best when they have a clear beginning, middle, and end. For some audience members though, the Internet, with its many links and cross-connections, is far less linear and may represent a better way of communicating.

---

[5]Gerald F. Else, trans., *Poetics/Aristotle* (Ann Arbor, MI: University of Michigan Press, 1970), 28, 30.

THE PREACHER'S MAXIM

If you've taken a writing or speaking course, the I-B-C or beginning-middle-end format will look familiar to you. For example, it resembles the recommendation put forth in the well-known preacher's maxim, which goes something like this: "First you tell 'em what you're gonna tell 'em, then you tell 'em, and then you tell 'em what you told 'em." It also looks much like the structure of a typical written essay.

The reason for repeating the message should be obvious—the more often people hear something, the more likely they are to remember it.

3. The beginning of a presentation should perform a variety of functions, most important of which is to explain why the speech is important and where it will be going.
4. The middle should stick to the topic as presented in the introduction.
5. The end should summarize the speech and make clear what, if anything, should or will happen next. Presenters often do a better job of organizing if they start by writing out the conclusion—once you understand how the presentation ends, it is much easier to develop the body and the introduction.

Even long speeches with many sections covering diverse subjects should retain a simple three-part structure that shapes the entire speech. This is true whether you are doing a five-minute overview at a staff meeting or presenting a ninety-minute detailed report on a year-long project.

Organizing with the I-B-C Format

Although all three parts of the I-B-C format contain important information for the audience, some sections receive more attention than others because of the way that people listen. Compare the parts of the I-B-C pattern and you will find that levels of audience interest vary greatly:

- **Introduction.** For a few brief moments you will have your listeners' rapt attention at the outset of the speech. At that point, they are most ready to be informed, persuaded, entertained, or instructed. Use this opportunity to create a good first impression—don't squander the chance.
- **Body.** Audience members tend to drift in and out during the body of a presentation; as you go along, their attention levels fall and rise. You may become frustrated when you realize your audience's level of attentiveness is going up and down in the middle of your talk. They are not

being disrespectful; they are merely processing what you say as you say it. It's not that your listeners aren't listening during the middle of the speech; it's just that they are absorbing what you say as you say it and so tend not to be as focused as they are at the beginning and end. This suggests that you will need to use rhetorical devices, such as changes in vocal tone or pauses in the speech, to keep their attention.

■ **Conclusion.** Your last chance to have an impact on your audience comes with the conclusion, and that last thing they hear is what they are most likely to remember. As they sense that you are coming to the end, audience members tune in to grasp words that summarize what they are supposed to have heard and motivate them to act. On a more basic level, they may be hoping for an inspiring close because they are tired of sitting or because they have many other things to do. In any case, give them a solid closing they will remember—don't just stop abruptly.

These different levels of audience interest should govern every aspect of your preparation for a speech. When you think about it, the most logical way to develop this kind of presentation is to start by writing the conclusion, then flesh out the body, and finish with the introduction.

Rough Out the Conclusion First

If you develop the conclusion first, you can arrange the rest of the presentation to lead up to that most important part. Clearly, the conclusion will have the most impact on the audience and must therefore be carefully designed to summarize all that has come before. Once you have a rough idea of how you want the presentation to end, you can write your conclusion. Assuming you have determined what your purpose is and have some ideas about how to achieve it, you can create a conclusion that will fit what you have been talking about. If you want your audience to act, feel, or think in a particular way, make your conclusion a powerful call for action and use it to reinforce the importance of what you are asking them to do.

After you are done with the conclusion, you can work on the body. To write the body of the presentation, use your outline and develop all of your main points in such a way that they lead into the conclusion. Each point should support what you plan to say at the end of your talk.

Once you know what happens in the body of the presentation, you can write an introduction that introduces your topic, forecasts where you are going with your talk, and explains how you plan to get there.

Don't Lose Sight of Your Main Objective

Focus on meeting the needs of the audience and remember the main rule governing every aspect of speech preparation and delivery: Speak for your listeners, not for yourself. If you always remember this rule, you will design introduction, body, and conclusion sections that fit together and accomplish your purpose.

HAVING SOMEONE INTRODUCE YOU TO THE AUDIENCE

One way to give yourself a bit more time to relax before you start talking is to ask someone to introduce you to the audience. Some people find it embarrassing, but it can add a lot to your credibility if you can find someone to introduce you to the audience before you speak. Follow these guidelines:

- Whenever possible, pick someone who knows you. That person is more likely to make you seem warm, human, accessible, and likeable. If the person who introduces you knows you well, you might encourage him or her to help you get your message across to the audience by stressing the main points of your talk. However, don't let that person give so much away in the introduction that there is nothing new in your speech.
- If the person who introduces you doesn't know you, write out a short introduction for yourself. Print your name at the top with a guide to pronunciation. List two or three of your accomplishments; the best ones are those that add to your credibility. Unless you are in an academic setting, avoid long lists of publications and research projects.
- You might also provide the person who introduces you with an anecdote— a story that helps illuminate your background, that demonstrates your talents, and that offers insight into your motivations and personality.

Guidelines for Introductions

The opening of a presentation serves a number of different functions. Following is a list of functions your introduction should accomplish, along with suggestions for implementing each.

Grab and Keep Their Attention; Make Them Want to Listen to You

Start with some kind of audience "hook," but *not* a joke. Possible openings include:

- A rhetorical question
- A screening question
- An anecdote
- A game, puzzle, or activity
- A worksheet or pre-test
- A statistic
- A startling statement or prediction—this could be a quote from someone famous or well-respected
- A credibility statement

Start with something that will catch and hold your audience's attention, such as an interesting anecdote or an unusual fact.[6] Whenever possible, start with an activity that gets the audience involved in your topic right away. If you want to make a stronger impression on the audience, interactive presentations (dialogues) have higher retention rates than lectures (monologues). Get your audience involved early on and they will stay with you longer.

Don't start with a joke. Unless you are a professional comedian, you are almost certain to mess up the joke, ruin the punch line, or (worse still) needlessly offend some members of your audience. Above all, don't start with a definition from the dictionary. This was fine in grade school; in the academic world it is considered childish.

Show the Importance and Relevance of Your Speech

With your purpose stated, you next must show the importance and relevance of the speech and generate audience interest. The amount of effort you expend on this step depends on the answers to the following questions:

- Do listeners have a personal interest in the subject?
- Does the topic relate specifically to their jobs?
- Will the topic have an immediate effect on their lives?
- Will the topic have a long-term effect on their lives?

The more "no" answers there are to these questions, the more likely it is that you must work to show the importance of the topic and create interest in it. Even when you know readers have interest, however, you still should spend a little time reminding them of why they should listen.

For example, if you are speaking to a college class about a proposed increase in tuition, you might want to add to the students' already keen interest with a statement such as, "If the tuition increase is approved and you are currently a first-semester sophomore, you'll end up paying $3,200 more tuition before you graduate than you would pay under the current rules."

You can establish importance and create interest with the same techniques that have been used by speakers for centuries. Each is effective in the right context, but each can also seem out of place if used with the wrong group. As always, follow Rule 1: Speak for your listeners, not for yourself, by using techniques that fit your audience and context.

With that caution in mind, here are some examples of how you can begin a presentation by showing importance and creating interest:

---

[6]If you are stuck for an opening for your presentation, check out the Idea Bank website at *http://www.idea-bank.com/*. They can help you "improve" speeches, articles, and presentations by using material from a huge collection of quotations, anecdotes, book and speech excerpts, proverbs, and humor.

- **Statistic.** Enrollment at this university has increased 40 percent since January 2001, but the number of parking spaces for students has stayed the same.
- **Rhetorical question.** As you reflect on your experience at this university over the last year or two, have you noticed that it's getting harder to find parking when you arrive for class?
- **Anecdote.** When I first started driving to school here, it took me twenty minutes from the time I left home to the time I walked into class. This semester, I have to leave thirty-five minutes for commuting—and I still live in the same place! What's taking up my time now is how long I have to drive around looking for a parking space. Last week I spent twenty-five minutes looking for parking, which made me late for class. I'll bet you each have your own tale of woe about campus parking.
- **Startling statement.** Faculty parking lots seem to be at least one-third empty every time I've observed them, but student lots are crammed full most mornings and early afternoons. Clearly, we have a parking problem on this campus.
- **Prediction.** If this university is not able to respond to the problem of inadequate student parking, students may begin to transfer to other schools with better student services.
- **Credibility statement.** For the last three years I've worked as a technical assistant for Rhodes Architects, Inc. Having assisted with several campus facilities design projects, I've developed some observations about the parking situation on our own campus.

Choose a method for showing importance that best suits the audience, the topic, and the speech situation. For example, the "startling statement" in the previous list might hurt your cause if you are speaking to a group of faculty and attempting to enlist support for building additional student lots. Find the right strategy for your specific listeners. Avoid an approach that might confuse, alienate, or bore them.

Defining Your Purpose

Listeners want to know why you are speaking. You may think the purpose is obvious, especially if the seminar program states that you will be speaking on "Concerns of the Cuban-American Community in Miami." However, the purpose is never so obvious that it should be omitted, and here is why:

- Making the purpose clear gives you a quick opportunity to establish a "comfort zone" for the audience. There will be no confusion as to what you intend to say and fewer complaints that you failed to meet audience expectations. Moreover, this is your best opportunity to make your topic relevant—to show your audience the connection between your topic and their needs, values, interests, and lives.

- Even if they've been told by others why you are speaking—as in a printed program or spoken introduction—listeners like to hear the purpose stated by you, the speaker. Their level of comfort expands as you confirm with your words what they have been told or what they have read.
- The purpose statement sets a clear expectation and helps establish a good relationship with listeners, who now have the first criterion by which to judge the success of your speech.

Some speakers wonder just how obvious a purpose statement should be. Although it's a bit boring to do so, there's nothing wrong with beginning your speech with the phrase, "The purpose of this speech is to . . ." A good introduction defines the scope and depth of your talk while providing a mental framework into which the audience can insert details covered in the body of the presentation.

EXAMPLES OF PRESENTATION PURPOSE STATEMENTS

In this example, the title of the presentation is "Concerns of the Cuban-American Community in Miami." These examples show several alternatives for phrasing the purpose.

- The purpose of this presentation is to describe diverse concerns of the Cuban-American community in Miami, Florida. (This is little more than a restatement of the title of the presentation.)
- This speech will describe how the Cuban-American community in Miami sees itself in relation to the other ethnic groups in the city. (This broad statement goes a little beyond what is in the title.)
- In this speech I will describe three major concerns of the Cuban-American community in Miami, Florida. (The scope of talk is limited to three major concerns. These should be enumerated in the next sentence.)
- The Cuban-American community in Miami has many concerns related to its traditional cultural background. Just what are these major concerns? (The focus will be on traditional Cuban-American culture.)
- Have you ever wondered what kinds of concerns have developed within the Cuban-American community in Miami? (This is a rhetorical question, which the speaker will answer.)

These examples give options for either direct or subtle statements of purpose. However, remember not to be *too* subtle about the purpose of the presentation. The listener should not have to work hard to find it.

Forecasting the Main Points of Your Speech

Forecast where your presentation is going by telling your audience or by showing them a list of your main points. You might want to turn this list into the handout referred to as a "bare-bones" outline. Leave a lot of space between items for audience members to take notes. Without an outline handout the audience has only your spoken words to navigate through the speech. And, even if they do have an outline, you must use obvious transitional statements to enable them to follow along.

Forecasting and transitional statements also help *you* as the speaker, because it will be easier for you to remember the wording and order of topics as you deliver the body of the speech. Use the following suggestions for writing a good forecast statement:

- Word points or topics exactly as they will be stated in the body of the speech so that listeners can follow easily.
- List points in the forecast statement in the same order they will appear in the body of the speech.
- If you have more than three main points to cover, consider using a visual aid to reinforce the points in your forecast statement.

Handled well, a forecast statement of main points in the end of the introduction becomes a "road map" for listeners to use throughout the presentation. It is especially helpful for listeners who have let their attention drift. They can find their way back when you mention points that you highlighted at the outset.

Although forecast statements don't give much room for creativity, here are a few variations:

Example: My presentation will examine three main reasons why we should begin a technical communication degree program on this campus: (1) jobs are available in the area, (2) salaries are high, and (3) costs for starting the program are low in comparison to those for the more technical programs on campus.

Example: Tanya Ruskin richly deserves the award of employee of the year. Within the company she led the Business Department's successful effort to install the new accounts-receivable system. And outside the firm, she played a key role in organizing a volunteer team to help build a home sponsored by Habitat for Humanity.

The introduction structure suggested here gives listeners a clear sense of direction. They know what topic you will cover, why it is important, and what points will be included. Although listeners prefer this level of clarity, some speakers have occasional doubts about such an "up-front" structure. They think it's more effective to "tease" listeners at the beginning by only hinting at the subject, which will become clear later.

Giving away the substance of your presentation when you start is the exact opposite of what you would do if you were presenting a drama or a comedy. Having events build from point to point until they reach a climax makes for better theater, but giving away the ending in the introduction is the professional way to do an informative or persuasive presentation.

Remember that listeners lack familiarity with the material. That's why you're giving the speech to them, not the other way around! Thus any ambiguities they face should be resolved as quickly as possible with an audience-centered introduction.

Establish Your Credibility

Establish your credibility in your introduction by listing your accomplishments, professional degrees, and affiliations. Set audience expectations in your introduction so they know what you will be presenting and at what level of sophistication. Remember, though, that reading a long list of publications and awards is not the best way to make a good first impression—it's boring and it sounds too much like bragging.

Give Audience a Chance to Get Used to Your Accent and Speech Patterns

Whether you realize it our not, most Americans have some kind of accent depending on the part of the country and culture they come from. For example, New Yorkers often find Texans unintelligible and vice versa. There may come a time when other forms of English are acceptable, but right now, if you want to be understood by the broadest segment of society, you must learn to speak in formal, standard American English.

Many speech communication problems, including accents, can be solved by speaking more slowly. Use your introduction to give your audience time to get used to your voice and to your accent while they learn more about what you will be presenting.

Guidelines for the Body

The body of the speech is the longest, most detailed section. Here you provide descriptive details for an informative speech, marshal your strongest arguments for a persuasive speech, or supply instructions and supporting material.

Show Relationships

Be prepared to have to explain the simplest and most obvious (to you) things if you have to. Prepare quick, simple definitions ahead of time and have them ready to pull out in case you need them. Use analogies to help your audience understand complex parts of your topic.

Beginners sometimes try to rush through a presentation as quickly as they can to get it over with. Use your time efficiently—don't rush and don't dawdle. If you find that your presentation is very short, use the additional time to go into more detail. You should overprepare—be ready for anything!

Limit the Number of Main Points

Don't try to pack too much into a presentation. You may have observed that your memory works best when items are grouped into just a few categories. The same holds true for your listeners, who are searching for convenient "handles" to aid them in understanding your speech. If you give them just a few categories into which to place details, they are more likely to remember material from the body of the speech. Many topics lend themselves to groupings of three, four, or five, but occasionally you will need to use more. The important thing is to work your main points into the time allotted for your talk without running over or confusing the audience.

In most cases, you already will have named the main points in the forecast part of the introduction. Therefore, you have given the audience an outline that is now reinforced by the structure of the body. The main points should, of course, appear in the same order in the body as they were presented in the introduction.

ORGANIZING THE BODY OF A PRESENTATION

Here are some examples of grouping ideas for different types of presentations:

- **Informative.** Five main skills you learned while completing an internship at the county courthouse
- **Persuasive.** Three problems that will be solved by adopting your proposal for a new media center at the college you attend
- **Occasional.** Four main contributions that support giving the "best student athlete" award to Morgan Smith, a graduating senior
- **Instructive.** Seven steps to completing a research proposal (Sometimes it is okay to have more than five.)

Choose the Most Appropriate Pattern

Each speech requires that you choose an organizational pattern for the body that best suits the particular speech. Here are some general patterns that can be used in any speech:

- Question/answer
- Cause/effect
- Problem/solution
- Comparison of options
- Chronological sequence of events
- Process description
- Classifying and describing parts of objects

Spend some time thinking about the kind of pattern that would be most appealing to your audience. For example, in an informative speech about your favorite job, the structure of the speech discussion could be based on (1) main job duties, (2) the sequence of tasks you confront during an average day, (3) the reasons why you chose this career, (4) lessons learned from the job, or (5) problems encountered during your time in the position. Your choice of a pattern would be determined by the purpose of the speech and needs of the audience.

Use a Mini-I-B-C Format Within Each Main Section

Just as the I-B-C format helps organize the entire speech, it also helps you reveal a sense of order in each of the main sections. Think of each section as a separate unit that, when assembled, produces a complete speech. Each of these mini-I-B-C units might do the following:

- **Introduction:** Briefly state the main point.
- **Body:** Give detailed support for that point.
- **Conclusion:** Summarize the point before moving on to the next one.

The main advantage of this "wheels within wheels" approach to organizing the body is that it provides listeners with a series of "mini-summaries," which are especially useful in more complex speeches.

Plan to Use Transitions Between Points

Transitions between the different parts of a talk are extremely important. They help the audience understand the relationship between the parts and thus gain a better understanding of the whole. If you lack imagination, you can always fall back on, "And then . . .," but there are other linguistic and visual devices that can help provide a smooth flow to your organizational plan. Here are a few transitional techniques:

- Use words at the beginning of each main point that indicate sequence (first, second, third . . .).
- Use words that indicate contrast (however, on the other hand).

- Pause between major points, look at your audience, and ask if there are any questions.
- Change to a new visual when you complete a section of your talk.
- Make distinct gestures at points of transition. For example, indicate past, present, and future by gesturing to one side for past, to the middle for present, and to the other side for future. Or you might deliberately walk from one side of the podium to the other to indicate a shift in topics.

Transitional devices help in several ways. First, they provide the auditory "glue" that makes the speech hang together as a coherent unit. Second, they present opportunities for you to recapture the attention of listeners who may have drifted off during the previous explanation.

It is sometimes a good idea to start a transition by recapping what has gone before. For example: "We've discussed the workings of the internal combustion engine and I have described how it can be combined with an electric motor to form a hybrid vehicle. Now let's look at the future of cars and the hydrogen fuel cell."

Planning your transitions ahead of time will improve the flow of the presentation.

Follow Every Abstraction with Specific Examples

Good speeches always depend on effective use of vivid, concrete information to support abstract points. Examples, anecdotes, stories, analogies, and illustrations will help drive home your points. Such details keep listeners engaged and help them recollect what you said.

Here are some examples of abstract points, followed by specific support to back them up:

Abstract idea: You enjoy gardening.
Concrete support: You include a story about how your parents bought you gardening tools and gave you a plot to garden when you were 10.

Abstract idea: Fish oil prevents heart disease.
Concrete support: You refer to the Eskimo-Aleut culture where salmon is a major part of the diet and where studies show heart disease is rare.

Abstract idea: Your college roommate had a strange sense of humor.
Concrete support: He once filled your car with turkey feathers before you were about to leave for a formal dance with your date.

As with writing, speeches are best received and remembered when they include a liberal application of vivid supporting examples.

Here are a few other points to remember about organizing the body of your talk:

- Stick to your plan—don't change things once you start talking or you will surely mess up.

- If you see puzzled looks in the audience, stop and ask if there are any questions.
- Don't rush. If you worked out your pacing in practice, you will be fine. Keep an eye on the clock, but don't change things in the middle of your presentation.

Guidelines for Conclusions

Conclusions are what the audience will remember best, so take the time to plan them very carefully. Unlike the rest of your speech, you might consider memorizing your conclusion so that it comes out sounding exactly the way you want it to.

A good conclusion ends with conviction:

- Make it clear that you are coming to the end. Try to finish on time or early, if possible (audiences always appreciate that).
- Have a solid conclusion that leaves your audience thinking about your message when they leave.
- Ask for questions if there is time. **Always repeat questions back so the whole audience can hear them before you answer them.** Prepare for tough questions ahead of time.
- Conclude again after the last question. Offer to answer complicated questions personally after the presentation and give out your business card, phone number, address, and e-mail so audience members can contact you if they think of a question later. This approach makes you look very professional.

Every speech conclusion has the same main objective: to summarize the speech and give the audience something to remember. There is no standard pattern of organization for accomplishing this objective, but the following sections outline general guidelines.

Restate the Main Points

Repeating the main points satisfies the last part of the preacher's maxim mentioned earlier: "Then you tell 'em what you told 'em." Audience members need to be reminded of your main ideas because they probably have no written list in their hands. Of course, you can reinforce this summary by a graphical representation of the list, if you choose to include a visual at this point.

Indicate What Happens Next

Conclude with a call for action, if appropriate. "Action statements" are appropriate for many speeches, especially those that aim to persuade. Following are some action statements that would fit in these types of speeches:

- **Informative speech.** Now that you know more about the World Wildlife Fund, here's how to contact the local chapter if you'd like to help with a project in your community.
- **Persuasive speech.** If you agree that more of our transportation tax should be spent on a rail system and less on highways, consider showing up at a hearing being held at the courthouse next Thursday.
- **Occasional speech.** Given a lifetime of support that Gordon Rice provided the YMCA, I hope you will consider continuing his work by making a financial contribution to the organization.
- **Instructional speech.** If you follow the basic maintenance steps I have just described, you will never again have to wonder if your bicycle is getting the care it needs to give you many years of safe riding.

Add a Personal Note

The very end of the speech gives you a moment to back out of the structured setting and finish with a personal comment, when appropriate. One example might be the kind of appeal to action mentioned in the previous guideline. Another might be a quick story, favorite quotation, or other device that reveals your own "humanity."

WHEN THE NEWS IS BAD

It sometimes happens that the purpose of a presentation is to break bad news or to explain something that is scary, painful, or unpleasant. In those situations, no amount of sugar coating is going to make your message a positive one.

All you can really do when the news is bad is to try to leave your audience with a bit of hope (and there are times you can't even do that). Be gentle.

There may be things you can say that will make the situation less frightening and less stressful, so do the best you can. Be prepared to have to answer the same questions more than once—audience members may need to hear you repeat your answers several times before they can accept the reality of the situation. Be patient.

When people are hurt and afraid, they can sometimes find comfort in talking it out. If you are comfortable with this, make your formal talk extremely short and use the bulk of your allotted time allowing audience members to ask questions, discussing their responses to your topic, and finding resources to keep them going.

Always Conclude Within the Time Limit

There is nothing else you can do in a presentation that will have the same positive impact on your audience as finishing on time (or even a bit early). In an age in which time is so precious, showing that you respect other people's time is guaranteed to make them think well of you.

End on a Respectful Note

Thank your audience for their attention and thank the person or group that invited you to speak. Offer to answer questions. (See Chapter 8.) After you are done taking questions, do a shorter, simpler version of your conclusion.

5.3 FINAL PREPARATIONS

Two more steps precede delivery of the speech and are essential for your success. First, choose the form in which you write presenter's notes to be used during the speech. Second, you must practice, practice, and practice some more. This section offers suggestions on completing both tasks.

Using Presenter's Notes

The best speeches are usually extemporaneous, in which the speaker shows great familiarity with the material and uses notes for reference. Here are some guidelines for making and using a reliable set of presenter's notes.

Guidelines for Making and Using Presenter's Notes

Don't Wing It

Beginners should always have a set of presenter's notes with them, even if they don't need to refer to them. Having a set of notes with you increases your chances for success.

Until you master the art of public speaking, you should avoid "winging it"—speaking without notes. Working "without a net" can result in disasters:

- If you lose your place, it is gone forever.
- If you are interrupted by a question, you will have trouble getting back on track.

Some people say they prefer the free-wheeling question and answer session to the organized presentation. They are more comfortable when given a chance to come up with a speech spontaneously than if they have time to prepare. The truth is that if you want to be a competent presenter, you will have to master both spontaneous and rehearsed speaking formats. You have much better odds of giving a good speech if you have a set of notes with you.

Avoid Reading a Speech or Reciting from Memory

In the first case you are ignoring your listeners, and in the second you risk being viewed as an automaton who is speaking mindlessly. Even if you know your speech by heart, it's best to look down at your notes occasionally to appear more natural in your delivery.

Besides helping you to appear more natural, extemporaneous speaking allows you to adjust phrasing and emphasis to improve delivery. In this way you are not locked into specific phrasing that you have memorized or written out word for word.

Pick a Format for Presenter's Notes That You Will Be Comfortable Using

There is no one right way to make presenter's notes. Some people prefer note cards, others prefer sheets of 8½″ × 11″ paper, while others print out a set of visuals and use those as an outline. Here are some methods for formatting a set of presenter's notes to be used during your speech.

Note Cards

The most traditional format for presenter's notes are 3 × 5, 4 × 6, or 5 × 7 note cards. Their main advantages are (1) readability, with just one or two points on each card, (2) convenience, in that they are easy to carry and store in a pocket or purse, (3) ease of revision, in that you can add, delete, or change the order of cards, and (4) mobility, in that you can easily hold them in your hand as you move beyond the lectern. It is difficult, but not impossible, to create a set of note cards on a word processor.[7]

There are also some disadvantages. For example, an inexperienced speaker may cause a distraction by flipping through cards noisily or waving them about.

Outline on 8½″ × 11″ Sheets of Paper

Some speakers prefer having an outline on 8½″ × 11″ paper because it allows them to see the complete speech on one or two sheets of paper. The standard outline format—with its points arranged in an indented hierarchy—gives the eye an easy reference point during the speech. The 8½″ × 11″ outline also assists you in avoiding the distraction of flipping note cards. However, reliance on sheets of paper makes it more difficult to move away from a lectern.

Presenter's Notes from PowerPoint Visuals

The PowerPoint **Notes view** lets you make notes for yourself as a presenter. You can use it to write down notes to yourself about when to pause, sip water, catch your breath, ask if there are questions, or give examples and other information to the audience. To be able to use the notes you make on the

---

[7]To create a set of note cards with Microsoft Word, pull down the **Tools** menu and select **Labels . . . .** Click on **Options** and select **Avery standard.** Then choose an Avery product number from the list shown: #5388 index cards (3 × 5) or #5389 post cards (4 × 6).

Notes pane during a presentation, you must print them out on paper and have them in front of you when you give your talk, because you won't be able to see them on the screen.

To print visuals on the same pages as your notes, you must add the notes in **Notes Page** view. Unlike **Normal, Outline, Slide,** and **Slide Sorter,** there is no **Notes** button at the bottom left of the PowerPoint window. In order to see and edit the **Notes Page,** you must pull down the **View** menu and click on **Notes Page.**

You can format your **Notes Pages** by pulling down the **View** menu, clicking on **Master ▶** and selecting **Notes Master.** The default format shows an $8\frac{1}{2}'' \times 11''$ page with a small version of each slide at the top of the page and room for a half-page of notes in outline form.

Options for Practice

Some speakers work hard on almost every stage of their speech but then fail to practice enough. Considering the importance of speeches to your personal and professional life, the extra time spent practicing is certainly warranted and usually pays off. Practice distinguishes superior presentations from mediocre ones. It also helps to eliminate the nervousness that most speakers feel at one time or another.

Some techniques for practicing your presentation are listed here:

Practice Before a Mirror

This old-fashioned approach allows you to hear and see yourself in action. The drawback, of course, is that it is difficult to evaluate your own performance while you are speaking. Nevertheless, such run-throughs can definitely make you more comfortable with the material.

Use an Audiotape

If have access to a tape recorder, this approach is quite practical. You can practice and listen to how you sound almost anywhere. Although taping a presentation will not improve gestures, it will help you discover and eliminate verbal distractions, such as filler words (*uhhhh, um, ya know*).

Use a Live Audience

Getting one or more of your colleagues, friends, or family members to observe your presentation can provide the kinds of responses that approximate those of a real audience. In setting up this type of practice session, however, be certain that your observers understand the criteria for a good presentation and are prepared to give an honest, forthright critique.

Use a Videotape

Video allows you to see and hear yourself as others do. At first it can be a chilling experience, but you will soon get over the awkwardness of seeing and hearing yourself on tape. Video also lets you see what your visuals will

LET YOUR COMPUTER HELP YOU PRACTICE

People have different ways of learning and different ways of recalling what they have learned. Most people are visual learners—they learn by seeing an image, reading text, or studying a graphic. Others are kinesthetic learners—they learn by touching, feeling, handling, and manipulating objects. A few are auditory learners, who learn by hearing things spoken by others.

If you are a predominantly "visual" person, you might want to use your computer to help you understand how your presentation will sound to an audience. Here are some ways of using a computer to help you practice your presentation:

- **Recording your voice.** Many modern computers come equipped with software for recording sounds and voices. Use your computer as a tape recorder to record and play back your voice.
- **Recording narration.** PowerPoint lets you record your words as a narration while you scroll through your visuals to accompany your slide show (pull down the **Slide Show** menu and click on **Record Narration**).
- **Text-to-speech.** Take advantage of the technology that is available to convert written words into spoken English. This "text-to-speech" software can enable your computer to "read" text aloud so you can hear how it sounds. While this doesn't quite compare to having a real person speaking, it can be very useful if you want to evaluate the flow and organization of a script.
- Apple computers come with text-to-speech software (pull down the **Apple** menu and click on **System Preferences,** then **Speech,** then **Spoken User Interface**).
- You can buy text-to-speech software for Windows computers. Many different software packages are available on the Internet.

look like from the audience's point of view. You can review the tape by yourself, but it can be even more effective if you watch with a qualified observer—someone who can help you identify and eliminate problems with posture, eye contact, vocal patterns, and gestures.

ENCOURAGEMENT

Do you want the audience to think, feel, or behave differently when you are done speaking? If you can focus on what you want to happen at the end of your talk, it will be much easier to develop your outline. Polish your conclusion until it shines and it will be a lot easier to organize the rest of your presentation.

TAKING IT TO THE NEXT LEVEL

Politicians, lawyers, and businesspeople who have to speak for a living rely on research assistants and hire speechwriters to produce their speeches. If it is critical that your talk be backed by the latest research findings, work with a professional librarian to make sure you have all the information you need and hire a writer to help you put your speech together.

EXERCISES

1. Write out a process description that pertains to your field. Turn the description into an outline for a presentation.

2. Each member of the class is to deliver a short impromptu talk on a topic selected by the instructor

3. Pick a simple process—how to make spaghetti, for example. Go around the room and have each person describe the next step in the process. Discuss which steps were omitted and which were the most difficult to describe in words.

4. Consider the following announcements of presentations How would you describe the most likely audience for each of these talks? Which appear to be more appropriate for the general public? Which are more appropriate for audience member with advanced degrees?

   - Chernobyl and Its Consequences: A look at the 1986 nuclear disaster from multiple perspectives, incorporating scientific and environmental approaches, as well as the literary and cultural presence of this defining event

   - Endangered Species Advocacy: Franklin's Ground Squirrel, a Case Study

   - CO_2 and Biodiversity Effects on Phenotypic Patterns of Selection for Two Prairie Perennials: *Lespedeza capitata* and *Schizachyrium scoparium*

   - Rotational Grazing of Scottish Highland Cattle for Shrub Management in Oak Savanna Restoration

   - Seminar on Multi-Scalar and High-Repetition-Rate Laser Diagnostics for Studies of Turbulent Combustion

   - Season's Eatings: How to Enjoy Thanksgiving and the Winter Holidays Without Gaining Weight

6 Illustrating Your Presentation

6.1 USE VISUAL AIDS

More than ever before, listeners expect good graphics during oral presentations. Graphics can help transform the words of your speech into true communication with the audience. The key word here is *help*. Too many speakers have decided that fancy graphics can substitute for features that always have been, and always will be, the hallmarks of an effective presentation—clear organization, enthusiastic delivery, and solid content.

Studies have shown that people have different ways of acquiring and recalling information. Most American adults (66 to 75 percent) are visual learners; their preferred way of getting information is by sight. Most of the rest are auditory learners; they prefer to hear information. A much smaller group is kinesthetic; they acquire information by touching and feeling objects. This has important implications for presenting: You will reach a much bigger audience if you add visual aids to a speech. This chapter will describe how to make and use visual aids in presentations.

6.2 MAKING GRAPHICS

Many large organizations today require employees to use specific templates and layouts if they are making their own visuals. Others have given the job of preparing presentation visuals to their graphic arts departments. In keeping with modern trends in "branding" and marketing, presentation visuals are seen as an expression of the organization's identity. Thus, every effort is made to achieve a level of consistency in the layout, design, and terminology used in these visuals so that they look slick and professional. If your organization specifies a template, you may have no choice but to use it.

Unfortunately, having graphic artists design visuals for presentations often leaves the technical expert who will actually be using the visuals out of the loop. If your organization requires you to have your visuals designed by an artist, you will need to work with that person and provide him or her with a clear outline of what you want to talk about. Needless to say, always preview and proofread visuals, especially if they are designed by someone else.

Fancy graphics cannot disguise an otherwise mediocre presentation, but they can enhance one that is already good. Graphics create a powerful effect on your audience if used skillfully. Let's begin by reviewing some basic guidelines that will help you use them well.

General Guidelines for Presentation Graphics

The best all-around suggestion for graphics is to view them as an integral part of the speech. This level of coordination can only be achieved if you plan graphics in the same way you plan your text—long before the speech is given. Think of text and pictures as a "team."

Plan Where to Use Visuals While Writing Your Outline

Actually, consideration of graphics should occur during your first musings about the speech. Graphics prepared as an afterthought usually look "tacked on." Plan them while you organize your talk so that the presentation will seem fluid.

This need for advance planning is especially important if you have to rely on others to prepare your visuals. Professionals need some lead time to do their best work. They can often provide helpful insights about what visuals will enhance your presentation—if you consult them early enough and if you make them part of your presentation team. Never put yourself in the position of having to apologize for the quality of your graphic material.

Discover Listener Preferences and Expectations

Some audience expectations come from years of tradition. For example:

- Math lectures are always given on a chalkboard, by a presenter working without notes.
- Art historians darken the room and use 35mm slides to illustrate their talks.
- Archæology presenters traditionally use a pair of 35mm slide projectors.

No matter what your field, your audience will have certain expectations of how you will use graphics and you choose a different format at your peril.

The most common setup used today is LED projection equipment connected to laptop computers or multiple slide projectors. However, some audiences prefer simple graphics—such as chalkboards, white boards, or flip charts—and the venue may not be equipped to handle anything more sophisticated. Listeners rarely indicate their preferences even when you ask

GRAPHICS IN A DOCUMENT VS. THOSE IN A PRESENTATION

In Chapter 3 we discussed the differences between writing and speaking. Consider these key differences between graphics for a document and graphics for a presentation:

- Documents provide the reader with ample time to examine detailed graphics. Speeches, however, don't give listeners an opportunity for careful review.
- For a presentation, graphics are enlarged to fill a projection screen, which may be 10 to 25 times larger than the original. An image that looks fine on a page printed at 300 dots per inch will lose an enormous amount of resolution when enlarged and projected on a screen. Presentation graphics need to be large and legible.
- A person reading a document can take his or her time examining images and, the more detailed the image, the more information they can get from it. This is especially true for maps. Because presentation visuals are often displayed for less than a minute, they shouldn't have too much detail. If details are important, give audience members a printed handout that they can take home and study at their leisure.
- In a document, it is possible (although not always desirable) to place text vertically, horizontally, or at an angle, especially in tables. If you try this on presentation graphics, your audience will have a lot of trouble reading them.
- Most people are familiar with standard methods of citing sources and adding notes to documents. Presentation graphics should show source citations, but there is no standard way of footnoting visuals, and some audience members may be frustrated or confused.

them, because they rely on the presenter to find the best format for illustrating a talk. It is your job to determine the best way to communicate information to the audience—two-dimensional (tables, bar charts, maps, photographs, or drawings), three-dimensional (models, props, live demonstrations), or animated (video, computer animation).

If you are unsure what is expected of you, contact key members of the audience and the event coordinator ahead of time to clarify the following points:

- Ask for information about the room in which you will be speaking and any peculiarities related to setting up equipment. Visit the venue yourself, if you can.

- Request a setting that allows you to make best use of the graphics preferences of your audience.
- Specify the kind of equipment you will need for your visuals.
- Choose graphics that best fit the constraints of the room (lighting, wall space, and chair configuration can influence your selection).

Keep Visuals Simple

Think of presentation visuals as though they were highway billboards—both need to communicate quickly and efficiently using a combination of text and graphics. Keep visuals simple, put complicated ideas and data in a handout, and give your audience time to read it and understand it.

Overly slick visual effects may not impress audience members. Some people prefer the simplicity of overhead transparencies with large, clear wording. If you use sophisticated technology, make sure it fits your context and audience.

Make Wording Brief and Visible

Some basic design guidelines apply whether you are presenting text on posters, overhead transparencies, or computer-aided graphics, such as PowerPoint:

- Use few words, emphasizing just one idea on each visual.
- Don't clutter up slides. Use lots of white space, perhaps as much as 60–70 percent per visual.
- The horizontal "landscape" format is preferable to the vertical "portrait," especially because it is the preferred default setting for much presentation software.
- Use type that is large enough to be visible in the room where you are presenting. Projection systems are all different, so you will have to experiment with different sizes of type to determine how large it needs to be. Generally, make type large enough so that there is one inch of height on the screen for each row of seats in the audience; for example, type on the screen should be 6" tall for an audience sitting six rows deep. When in doubt, make the type larger (and use less of it).
- Use both upper- and lowercase text—not ALL CAPS—because it is more legible.

Use Colors Carefully

Colors can add flair to visuals. Have a good reason for using color (such as the need to highlight three different bars on a graph with three distinct colors). Use easily seen colors, and be sure that text color contrasts with its background (for example, white on pale blue does not work well). Use a simple background and no more than three text colors in each visual (to avoid a confused effect). For variety, consider using white text on a black or dark background.

Practice Using Your Graphics

Include every graphic you plan to use in your practice sessions. This is a good reason to prepare graphics as you prepare text, rather than as an afterthought. Running through a final practice without graphics would be like doing a dress rehearsal for a play without costumes and props—you would be leaving out parts that require the greatest degree of timing and orchestration.

Check Spelling

It's always smart to proofread your text for typos and factual errors *before your presentation*. Don't embarrass yourself.

Make Graphics Work For You

Good graphics can make the job of presenter a lot easier. Bad graphics can slow everything down and confuse the audience. If a picture is indeed "worth a thousand words," every good graphic you use will save you lots of time and trouble in your presentation.

Make a Title Slide

Even if you decide not to use a lot of presentation graphics, you should still create a visual to be shown at the very beginning (or very near the beginning) of your talk—stating the title of your presentation and telling your audience your name. You might also include information about how people can get in touch with you if they have questions (or if they'd like to hire you to do another presentation!).

Make a Concluding Visual

This will let your audience know your presentation is over and that you are willing to take questions. Cite references for images and ideas right on the visuals as you show them. Include a list of resources at the end in case the audience wants to know more.

Guidelines for Using Numbers in Presentations

Using numbers can be challenging. Here are some guidelines to help you with using numbers.

Project Numbers on a Screen

It can be hard for audience members to follow or remember numbers, so it is a good idea to put them up on a visual. Complicated formulas and equations work better when they are projected and explained term-by-term to the audience. Many people suffer from "math anxiety," so give your audience examples and analogies to help them understand mathematical concepts.

USING MICROSOFT GRAPH WITH POWERPOINT

If you use PowerPoint, you can use a program called Microsoft Graph (Figure 6–1) to make different figures for presentation graphics. Examples of chart types you can create with Microsoft Graph include Area, Column, Bar, Line, Pie, Doughnut, Stock, XY (scatter), Bubble, Radar, Surface, Cone, Cylinder, and Pyramid. Here are descriptions of the different types of charts:

Column: A column chart shows data changes over a period of time or illustrates comparisons among items.

Bar: A bar chart illustrates comparisons among individual items.

Line: A line chart shows trends in data at equal intervals.

FIGURE 6–1
Microsoft Graph chart types

Pie: A pie chart shows the proportional size of items that make up a data series to the sum of the items

XY (scatter): An XY (or scatter) chart either shows the relationships among the numeric values in several data series or plots two groups of numbers as one series of *x, y* coordinates. This chart shows uneven intervals—or clusters—of data and is commonly used for scientific data. When you arrange your data, place *x* values in one row or column and then enter corresponding *y* values in the adjacent rows or columns.

Area: An area chart emphasizes the magnitude of change over time. By displaying the sum of the plotted values, an area chart also shows the relationship of parts to a whole.

Doughnut: This form of pie chart does not work well for presentation visuals.

Radar: In a radar chart, each category has its own value axis radiating from the center point. Lines connect all the values in the same series. A radar chart compares the aggregate values of a number of data series.

Surface: This format does not work well for presentation visuals.

Bubble: A bubble chart is a type of XY (scatter) chart. It does not work well for presentation visuals.

Stock: The high-low-close chart is often used to illustrate stock prices. This chart can also be used for scientific data, for example, to indicate temperature changes.

Cone, Cylinder, and Pyramid: These 3D data markers do not work well for presentation visuals.

Two terms you should know when working with these visuals are *trendlines* and *moving average.* Trendlines are used to graphically display trends in data and to analyze problems of prediction. Such analysis is also called *regression analysis.* By using regression analysis, you can extend a trendline in a chart beyond the actual data to predict future values. A moving average smoothes out fluctuations in data and shows the pattern or trend more clearly.

Distinguish Between Quantities, Percentages, Percentiles, and Averages

Technical experts work with all kinds of numbers, so it is important that you be clear about what figures mean.

- Percentages are always part of a whole.
- A percentile is a value on a scale from one to one hundred that indicates whether a distribution is above or below it. (A score at the third percentile means 3 percent of the other scores were lower; a score of ninety-third percentile means 93 percent of the other scores were lower.)
- Data points often represent an arithmetic mean (average) rather than an exact number. If you use data and statistics, make sure you understand and can explain how the data was collected, how large the sample was, and what the margin of error is.

In each case, you have an obligation to explain to your audience what the numbers you are using mean.

Use Both English and Metric for Measurements for American Audiences

Most of the rest of the world uses the metric system only, but when speaking to an American audience, you *must* use both English and metric measurements.

Don't Write Dates with Slashes

Americans use slashes to indicate month/day/year, but in many other parts of the world, slashes are not used at all or, if they are, they represent day/month/year. Thus 2/1/06 is the first day of February in the United States, but in other parts of the world, this designates January 2. Avoid writing dates with slashes. Instead, use one of these variations of month, date, and year:

- November 8, 2004
- 8 November 2004

Convert Dollar Amounts to Value in a Single Year (e.g., in 2000 dollars)

The U.S. Department of Labor Bureau of Labor Statistics website at *http://www.bls.gov/home.htm* has an Inflation Calculator that you can use to convert any dollar amount from one year to another. While no one can predict the future, this site can be helpful for comparing prices today with those of past years.

Use Scientific Notation to Eliminate Extra Zeros

Write very large or very small numbers using scientific notation. For example: You would write .0000000000000142 as 1.42×10^{-14}.

Define Symbols, Terms, and Variables

Don't assume that your audience knows what a symbol like Ω stands for. Make sure you define unfamiliar or unusual symbols clearly and carefully, and, if you have to use several symbols or terms, make up a handout that explains and defines them.

Limit Repeating Elements

You can leave out repeated elements—like %, \$, €, °C—as long as you indicate somewhere on the visual that amounts are percents, U.S. dollars, euros, or degrees Celsius. If every single item is in thousands of dollars, you can take out the extra zeros and add a note: "In thousands of dollars." Similarly, if every item is in pounds per square inch, you don't need to put the abbreviation "p.s.i." next to every item. The same holds true for dollar signs and other elements that may be repeated many times on one visual. This will reduce a lot of the clutter that can make data graphics hard to understand. Put a \$ symbol next to the first number and your audience will infer that the rest of the numbers are also in dollars.

Types of Visuals

Visuals can be classified as text, data graphics, or representational graphics. Figure 6–2 illustrates these categories and shows how the next section of this book is organized.

- **Text** can be in the form of lists or quotes. Some tables also use text.
- **Data graphics** are visuals designed to help audience members understand relationships between numbers. The two main types of data graphics are tables and charts.
- **Representational graphics** can be categorized as either pictorial or symbolic representations of reality. Pictorial representations include 2D and 3D models, photographs and videos, and drawings. Symbolic representations include maps, flowcharts, organization charts, or computer visualization and plotting images.

Deciding What Visuals to Use

It is helpful to categorize the information you want to put on visual aids before you decide how to display it. Table 6–1 shows just a few ideas for visuals.

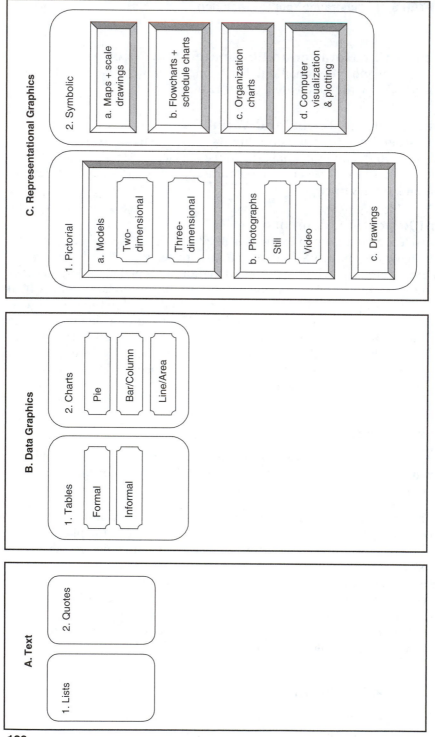

FIGURE 6–2

Types of graphics for illustrating presentations

TABLE 6–1
Presentation Visuals

To show:	Consider using these visuals:
People	Photographs or video
Places	Maps help audience understand *where;* photographs and video can show them *what.*
Objects	Best is to bring out the actual object. High-contrast photos are okay, but schematics, exploded view drawings, and 3D models are also good.
Processes	Flowcharts, schedule charts (Gantt, milestone, and PERT charts) are the best for symbolic representations. A short, well-made video is very good if you want to show the process in action.
Trends	Line graphs that show trends over time work well.
Precise numbers	Tables are best for this, but formal tables have so much detail that they are very difficult to use in presentations.
Changes over time	A series of photographs or instrument readings taken from one vantage point can work. Visualization graphics and animations are great, but can be costly and time-consuming.
Activities	Video or animations can show movement and changes in relationships over time and in three dimensions.
Parts of a whole	Pie charts are good for a small number of categories.
Comparing options	A table can show details of comparison; use bar or column graphs if exact numbers aren't important.
Correlation between variables	A scatter plot with trend line (regression) works well, as well as simple visualization and plotting graphics.
Organizational structure	Use an organization chart.
Very small objects	Place objects under an overhead document camera like the one shown in Figure 6–4 and zoom in to magnify.

6.3   *TEXT*

Never load up visuals with text. Text-heavy visuals tend to discourage audience members who may consider reading to be work. Aside from using text for descriptive captions on graphics, for callouts, and to show sources of borrowed material, you will be more successful if you limit your use of text in visuals to lists, quotations, and comparison tables (you will also be less likely to read your visuals word-for-word to the audience). Here are some guidelines for using text in presentation visuals:

- Stick to one or two typefaces. (Murphy's Law says your favorite font will not be available on the computer you will use to show your visuals.)
- Text size should be determined by the distance between the screen and the back of the audience—generally text should have one inch of height for each row of seats.
- Choose terminology carefully and use it consistently. Don't refer to an object as an "On Button" on one visual and as a "Control Switch" on another.
- Use callouts to label parts of images. Keep callouts very simple. Figure 6–3 shows the different kinds of callouts available with PowerPoint and Figure 6–4 demonstrates how callouts are used.
- Pictures are almost always better than text-based visuals. Consider replacing some text with images that will help your audience understand your topic. A good, clear picture, like the one of a document camera

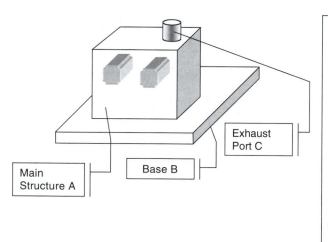

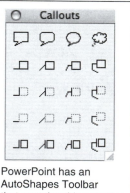

PowerPoint has an AutoShapes Toolbar that contains this selection of callouts, which can be used as text boxes.

FIGURE 6–3
Examples of callouts

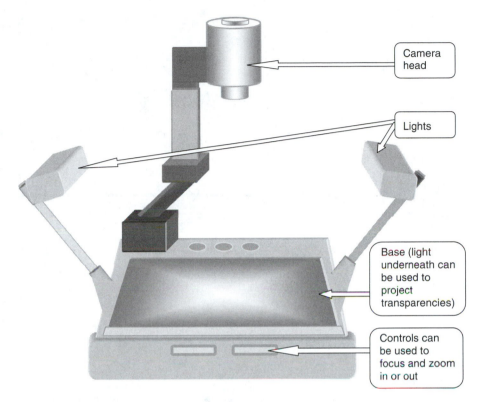

FIGURE 6–4
Overhead document camera

shown in Figure 6–4, can make it much easier for your audience to understand your topic.

- Never project a printed page as a visual unless you have time to let the audience read all of it. If it is important for the audience to see what a document looks like, use an overhead document camera and zoom in as much as you can. Overhead document cameras, like the one pictured in Figure 6–4, send a live video image to a projector.[1]

Lists

Lists make good visuals when they are simple and easy to understand. Figure 6–5 shows a list from the presentation outlined in Chapter 5. The number before the title corresponds to the presentation outline. Note the use of short

---

[1]Document cameras are also known as visualizers, visual presenters, digital presenters, and visual copy stands. Manufacturers of these devices include AverMedia, Canon, Elmo, Epson, Samsung, Toshiba, and WolfVision.

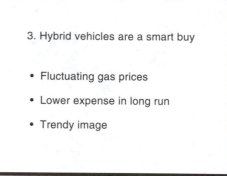

FIGURE 6–5
Example of a list as a presentation
visual

phrases rather than sentences. Here are some tips for using lists as presentation visuals:

- Use keywords and phrases; avoid sentences.
- Use parallel structure for list items.
- Use bullet lists when you don't want your audience to infer a rank order of items. Use numbers or letter items when you have a logical reason for ranking them.
- Many presenters have been taught a "66" rule for visuals: no more than six lines of text with no more than six words per line. This "rule" can and should be broken if circumstances dictate. Complex ideas may require more than six bullet points to explain. If you have more than six, group your points into categories and make those your list. If you have a longer list of relevant points, but don't plan on discussing them all in your time-limited presentation, put them into a handout for your audience to read at its leisure.

Quotations

If you want to display someone's exact words, type or paste the quotation into a visual and leave it up long enough for everyone to read. Figure 6–6 shows what a visual with a quotation looks like. (Actually, the example shown is a quote within a quote!) This example, which contains about seventy words, should be left up for at least a minute so the audience can read the entire thing. Note that the last part contains an unexpected twist that adds a touch of humor to the presentation.

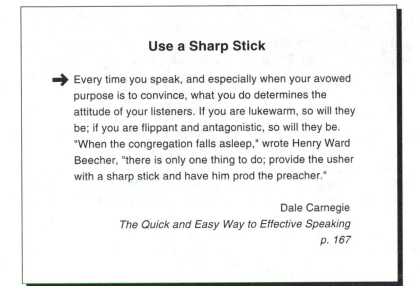

Use a Sharp Stick

➡ Every time you speak, and especially when your avowed purpose is to convince, what you do determines the attitude of your listeners. If you are lukewarm, so will they be; if you are flippant and antagonistic, so will they be. "When the congregation falls asleep," wrote Henry Ward Beecher, "there is only one thing to do; provide the usher with a sharp stick and have him prod the preacher."

Dale Carnegie
The Quick and Easy Way to Effective Speaking
p. 167

FIGURE 6–6
Example of a quotation in a presentation visual

6.4 DATA GRAPHICS

Data graphics show relationships between numbers. Each of the data graphics discussed here can enhance a presentation if used appropriately. While there is no "right" way to illustrate a presentation, some techniques are much less effective than others and can hurt a speaker's ability to communicate clearly. At the end of this chapter you will find a Rogues' Gallery of problem graphics, with examples of what can go wrong for each type of graphic discussed here.

The guidelines presented here transcend issues of technology. You don't have to use a computer to create these visual aids, but your presentation will come across as more professional if you do.

Tables

Tables present the audience with a grid (sometimes called a *matrix* or an *array*) containing raw data, usually in the form of precise numbers, but sometimes in the form of words. Tables serve two main purposes: They can show exact numbers and they can display comparisons. Any graphic that is not a table is called a *chart* in PowerPoint and a *figure* by technical writers.

Table cells display numbers or other data in a grid without performing mathematical operations. Tables are not spreadsheets: Spreadsheets, like those created with Microsoft Excel, are designed to make manipulation of numbers easy for accounting. In a spreadsheet, hidden formulas perform calculations on the visible data. Spreadsheets do not make good visual aids because your audience can't see those operations.

The following guidelines will help you design tables as visual aids:

Use Simple, Informal Tables When Possible

Tables are classified as either informal or formal. Informal tables show limited data arranged in either rows or columns, similar to a list. Formal tables contain complex data arranged in a grid, always with both horizontal rows and vertical columns. Informal tables usually are preferred for presentation visuals because they include only rows or columns, not both. Figure 6–7 shows an informal table that has (1) no table number or title, and (2) few, if any, headings for rows or columns. For complex data, and if all the data are important for your presentation, use formal tables.

When you use formal tables, do the following:

- Extract important data from the table and highlight them in the speech. Use color or boldface to make important information stand out from surrounding detail.

Table 2: Staffing for Alberta Project

Office where staff will come from:

San Francisco	42
St. Louis	31
Las Vegas	8
London	5
Total	86

Our project in Alberta, Canada, will involve engineers, technicians, and salespeople from three other offices.

FIGURE 6–7
Example of an informal table as a presentation visual

- Leave the table on the screen long enough for the audience to absorb important information. Bring it back if it is needed again.
- Do not use a formal table if most of the data will be ignored. It would be more of a distraction than an aid.
- Do not import a table that was designed for a document directly into a presentation visual. Tables in documents can be extremely detailed— designed to be studied slowly and carefully. They are often printed in the smallest type size available. The example shown in Figure 6–8 was derived from a much larger table. The items are arranged in order of decreasing percent increase, from 59 percent down to 40 percent. The number of jobs is given in thousands, which eliminates the need for lots of zeros. The meaning of the quartile rankings is explained in a note below the table.
- If you try to use a table with too many columns and rows as a presentation visual, the type will be so small that your audience will not be able to see the contents of the cells. When you use large formal tables for complex data, put those tables into a handout for your audience to study at their leisure.

Ten fastest growing health-related occupations, 2002-12

Occupation title	Change from 2002 to 2012		Quartile rank by 2002 median annual earnings*
	Number of jobs (in thousands)	Percent increase	
Medical assistants	215	59	3 low
Physician assistants	31	49	1 very high
Home health aides	279	48	4 very low
Medical records and health information technicians	69	47	3 low
Physical therapist aides	17	46	3 low
Physical therapist assistants	22	45	2 high
Dental hygienists	64	43	1 very high
Occupational therapist aides	4	43	3 low
Dental assistants	113	42	3 low
Personal and home care aides	246	40	4 very low

*Rankings based on quartiles using one-fourth of total employment to define each quartile. Earnings are for wage and salary workers.

Source: Occupational Projections and Training Data, Bulletin 2572 (Bureau of Labor Statistics).

FIGURE 6–8
Example of a formal table as a presentation visual

Use Plenty of White Space

Don't cram information into table cells. Used around and within tables, white space guides the eye through a table much better than do black borders. Avoid putting densely drawn black boxes around tables. Instead, leave more white space than you would normally leave around text and let it act as a frame.

Follow the Design Conventions for Tables

A formal table satisfies the overriding goal of being clear and self-contained. To achieve that objective in your tables, follow these guidelines:

- **Titles and numberings.** Title each formal table, and place title and number above the table. Number each table if the speech contains two or more tables.
- **Headings.** Create short, clear headings for all columns and rows.
- **Table items.** Place items in some logical order—most expensive to least, shortest to longest, oldest to newest. Large tables used in documents may have items listed in alphabetical order, but such tables would probably be too large for a presentation visual.
- **Abbreviations.** Include in the headings any necessary abbreviations or symbols, such as p.s.i. or %. Explain abbreviations and define terms if listeners need such assistance.
- **Numbers.** Use tables when you want to show your audience exact numbers. If exact numbers are not important, use a figure rather than a table. Use the decimal tab to align numbers on the units column or at the decimal (when shown). Use scientific notation to reduce decimal places. Take out zeros in thousands and millions. (Put a note on the visual so your audience isn't confused.)
- **Notes.** Place any explanatory notes either between the title and the table (if the notes are short) or at the bottom of the table.
- **Sources.** Place any source references on the visual and make reference to the source during your speech.
- **Caps.** Use uppercase and lowercase letters, rather than all caps.

Remember that the entire audience should be able to read all writing. If it is not readable remove it from the table during preparation and give the information (such as sources and notes) in your speech.

Pay Special Attention to Cost Data

Audiences prefer to have financial information placed in tabular form. Given the importance of such data, edit cost tables with great care. Devote extra attention to these issues:

- Placement of commas and decimals in costs
- Correct figure totals

Charts

Any information that doesn't fit into a table grid is called a *figure* or a *chart*. Other figures are called *maps, pictures, schematics, flowcharts,* and *photographs*. When creating a pie chart with Microsoft Graph for PowerPoint, start by clicking on the **Insert Chart** icon (or pull down the **Insert** menu, click on **Object** . . . and select **Microsoft Chart**). Depending on the version of PowerPoint you are using, the default may be a column or bar chart. To change to a different type of chart, pull down the **Chart** menu and click on **Chart Type** . . . as shown in Figure 6–1.

Pie Charts

Pie charts are only used to show approximate relationships between the parts of a whole and then only when you want the audience to be able to compare the parts to one another. Their simple circles with clear labels can sometimes be helpful, but the design simplicity of the circle means that pie charts are not the best choice when you need to show detailed information or changes over time.

To create a pie chart with Microsoft Graph for PowerPoint, click on **Chart Type** . . ., switch from the default to pie chart, and fill in the cells of a spreadsheet (like the one shown in Figure 6–9) with the names of the segments in the top row and the values of the segments in Row 1 below that.

The spreadsheet shown in Figure 6–9 will result in a pie chart that looks like Figure 6–10. Some of the defaults have been changed to make the chart easier to read. Labels for the segments replace the legend box. Solid color segments have been changed to patterns so they are easier to see, even in black and white. Following are basic guidelines for constructing pie charts.

Use No More than Six or Seven Segments

To make pie charts work well, limit the number of pie segments to no more than six or seven. In fact, the fewer segments the better. This approach lets the audience grasp major relationships without having to wade through the clutter of tiny divisions that are difficult to see.

Remember that 10 percent of all men are color-blind, so use patterns for fills and lines. Each segment of a pie chart should be a different color and pattern. In Figure 6–10, for example, the audience can readily see that most of the sales come from three offices in Georgia.

		A	B	C	D	E	F
		Savannah Off	Macon Office	Atlanta Offic	Mobile Office	Other Offices	
1	Percent	40	25	15	15	5	
2							
3							

Graph in PIE Sales by office.doc – Datasheet

FIGURE 6–9
Spreadsheet used to create a pie chart

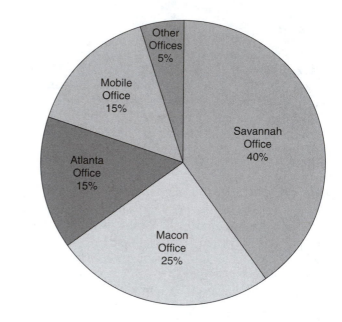

FIGURE 6–10
Pie chart made with
Microsoft Graph

Order Segments Logically

Pie charts should be oriented like a clock, with the first and largest segment starting at 12:00. Move clockwise from the largest to the smallest segment to provide a convenient organizing principle. Pie charts should never be organized alphabetically. *Note:* Some presentation software packages will not permit you to begin the pie at 12:00.

Use Pie Charts for Percentages and Money

Pie charts catch the observer's eye best when they represent items divisible by 100, as with percentages and dollars. Using the pie chart for money breakdowns is even more appropriate because of the coin-like shape of the chart. In every case, make sure your percentages or cents add up to 100.

Be Creative, But Keep It Simple and Avoid Clutter

Some presentation software programs automatically format pie charts with complex backgrounds and shading. Others turn pie charts into three-dimensional "hockey pucks" or pull the segments apart. These fancy bells and whistles may be fun to play with, but they are difficult to read and they distort the pieces of the pie. Avoid them when making pie charts.

Draw and Label Carefully

The most common pie chart errors are (1) segment sizes that do not correspond correctly to percentages or money amounts and (2) pie sizes that are too small to accommodate the information placed on them. Here are some suggestions for avoiding these mistakes:

- **Pie size.** Make sure the chart occupies enough of the page. Each visual should have no more than two pie charts, with each circle large enough to show up clearly when projected on a screen during your speech.
- **Labels.** Place labels either inside the pie or outside, depending on the number of segments, the number of segment labels, or the length of the labels. Choose the option that produces the cleanest-looking chart. As always, remember that all wording on a chart must be clearly readable by everyone in the audience.
- **Conversion of percentages.** If you draw a pie chart by hand, use a protractor. There are 360 degrees in a circle; so 1 percent of a pie equals 3.6 degrees. Thus, you can convert percentages or cents to degrees.

Remember that a pie chart does not reveal fine distinctions very well; it is best used for showing larger differences.

Bar/Column Charts

Bar (horizontal) and column (vertical) charts are useful for simple comparisons. Bar/column charts are easily recognized and are often found in newspapers and magazines. However, bar charts can accommodate more technical detail than pie charts without appearing cluttered. Comparisons are provided by means of two or more bars running either horizontally (bars) or vertically (columns). The column chart in Figure 6–11 shows how effective

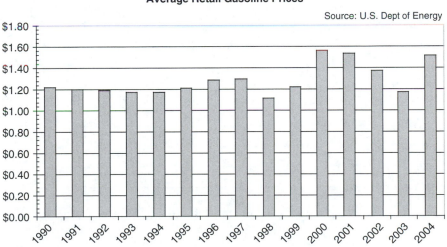

FIGURE 6–11
Column chart

this format is for displaying information. Note that the last column is clearly labeled as an estimate, so as not to mislead the audience.

Stock charts are a form of column chart where the column is replaced by a line and tick marks indicating the high, low, and closing price of a stock. Stock charts can also be used for other kinds of data, as shown in Figure 6–12, which has the same information contained in Figure 6–11 with additional information about the types of gasoline. The "high" has been replaced by premium, high-octane gas and the "low" represents the price of regular gas.

Follow these guidelines to create effective bar charts:

Limit the Number of Bars

Although bar charts can show more information than pie charts, both types of illustrations have their limits. Bar charts begin to break down when there are too many bars. The maximum number of bars can vary according to chart size, of course. However, reducing a bar chart's number of bars enhances its impact on the audience.

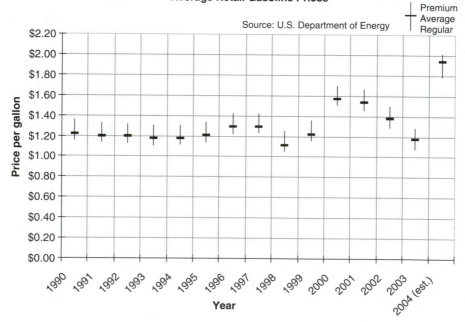

FIGURE 6–12
Stock chart used to show gas price changes

Show Comparisons Clearly

The length of the bar or column illustrates the values being compared. A good bar or column chart should convey an immediate visual impact.

- Bar lengths should be varied enough to show comparisons quickly and clearly. Never use a logarithmic scale with bar or column charts.
- Avoid using bars that are too similar or too different in length, for then the audience must study the chart for a long time before grasping its meaning.
- When there is a big discrepancy between data points, some presenters resort to the dubious technique of inserting "break lines" (two parallel lines) to reflect breaks in scale. Although this approach at least points out the breaks, it is deceptive.

Keep Bar Widths Equal and Adjust Space Between Bars Carefully

The length of the bar or column illustrates the values being compared. Distorted bar/column charts mislead people into comparing the *area* of the bar or column rather than the *length.* Therefore, while the length of a bar or column will vary, bar width must remain constant.

Arrange Bars or Columns in Some Logical Order

The arrangement of bars reveals much of the chart's meaning to the audience. Here are two common approaches:

- **Sequential.** Used when the progress of the bars shows trends—for example, the increasing number of environmental projects in the last five years
- **Ascending or descending order.** Used when you want to make a point by the rising or falling of the bars—for example, total sales of a firm's six offices for 2004, from lowest total to highest total

Be Creative, But Be Careful

Stacking columns may work in documents in which readers can take their time studying, but they are usually far too complex for presentation visuals. The example shown in Figure 6–13 is effective because it has only two columns and they are very similar. These stacked columns help explain that, even though the price of gasoline increased by more than 15 percent between 2002 and 2003, the relative size of costs changed very little.

As a general rule, avoid stacked columns or bars and don't use fancy 3D effects that hide data. If you must show two or more trends at the same time, use simple line charts, rather than bar/column charts. In fact, it is always a good idea to keep visuals simple. Remember, if they don't make it easier for audience members to understand what you are talking about, they may actually hurt your presentation.

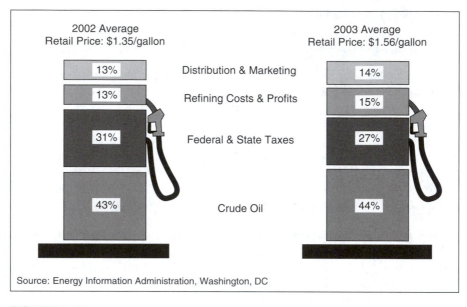

FIGURE 6–13
Stacked column chart

Line (Alpine) Charts

Line charts (sometimes called *alpine charts* because of their resemblance to a mountain range) are a common graphic in speeches and documents. They are useful for showing trends, especially if the precise numbers are not as important as the overall pattern. And they are important for showing the relationship between two variables. Almost every newspaper contains a few line charts covering topics such as stock trends, car prices, or weather. More than other graphics, line charts telegraph complex trends immediately.

Figure 6–14 shows how line graphs are used to plot changes over time. Note the absence of vertical lines and that, even though the graph plots weekly data points, the *x* axis only has dates for every other month.

Line charts work by using vertical and horizontal axes to compare two different variables. The vertical (Value or *y*-) axis usually plots the dependent variable; the horizontal (Category or *x*-) axis usually plots the independent variable. (The dependent variable is affected by changes in the independent variable.) Lines then connect points that have been plotted on the chart.

Use Line Charts to Show Trends over Time

The audience is influenced by the direction and angle of a chart's line, so take advantage of this persuasive potential. For example, a speaker who wants to show the feasibility of adopting a new medical plan for an organization can

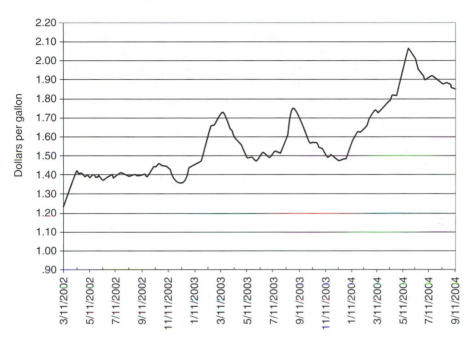

Weekly U.S. Retail Gasoline Prices, Regular Grade

Source: Energy Information Administration, Washington, DC

FIGURE 6–14
Line graph

include a line chart that gives immediate emphasis to the most important is-sue—the effect the new plan will have on stabilizing or lowering the firm's medical costs.

Use Line Charts with Care
Given their strong impact, line charts can be especially useful as attention grabbers. Consider using them (1) at the beginning of your speech to engage interest, (2) at the beginning of sections that describe trends, and (3) in your conclusion to reinforce a major point of the speech. Because they are so ef-fective at communicating information, resist the temptation to pack your presentation with too many line charts.

Strive for Accuracy and Clarity
Like bar charts, line charts can be misused or poorly constructed. Be sure the line or lines on the graph accurately reflect the data from which you have drawn. Also, select a scale that does not mislead readers with visual gimmicks.

There are certain conventions you should follow when creating a line chart:

- Time starts at the left and runs along the *x*-axis to the right.
- Never put two different scales along the same axis when designing a presentation graphic. This might work with line charts in documents, but it is too confusing for an image that will only be projected for a few seconds.
- Time lines should go back far enough in time so that the graphic isn't deceptive.
- Start all scales from zero to eliminate the possible confusion of breaks in amounts. Unless there is some strong, compelling reason to do otherwise, both the *x*- and *y*-axes should start at zero.
- Select a vertical-to-horizontal ratio for axis lengths that is pleasing to the eye (three vertical to four horizontal is common).
- You want the information to pop out of the graphic, so number and limit the thickness of gridlines or make them a lighter color. Use bold lines that are easy to see to show the trends and use very few, very light grid lines. Make chart lines as thick as, or thicker than, the grid lines. When it comes to grids, the less "ink," the better.
- Line charts can be turned into area graphs, where the area below the line is filled in. Do this only when there is a single line and where it will make the chart more readable.

Do Not Place Numbers on Line Charts

Line charts derive their main effect from the simplicity of lines that show trends. Avoid cluttering the chart with a lot of numbers that only detract from the visual impact.

Do Not Use Multiple Lines or Log Scales on Line Charts

Like bar charts, line charts in a document can show multiple trends. But in a presentation graphic, if you place too many lines on one chart, you risk confusing the audience with too much data. For the same reason, never use a logarithmic scale with a line chart.

Use Patterned Lines

If you show more than one line on a graph, make them different colors and different patterns in case some audience members are color-blind.

6.5   MISUSE OF DATA GRAPHICS

Technology has revolutionized the world of graphics by placing sophisticated tools in the hands of many users. Yet this largely positive event has its dark side. You may have seen some graphics that—in spite of their professional appearance—distort data and misinform the audience. The previous sections of this chapter have established principles and guidelines to help you avoid such distortions and misinformation. This next section shows what can happen to graphics when sound design principles are not applied. It also offers suggestions for correctly employing that increasingly popular form of speech graphics—PowerPoint.

Description of the Graphics Problem

The popular media provide one window into the problem of faulty graphics. One observer used newspaper reports about the October 19, 1987, stock market plunge as an indication of the problem. Writing in *Aldus Magazine*, Daryl Moen noted that 60 percent of U.S. newspapers included charts and other graphics about the market drop the day after it occurred. Moen's study revealed that one out of eight had data errors, and one out of three distorted the facts with visual effects. That startling statistic suggests that faulty illustrations are a genuine problem.[2]

Edward R. Tufte coined the term *chartjunk* to describe ornate data graphics that fail to communicate information. Tufte analyzes graphics problems in more detail in his excellent work, *The Visual Display of Quantitative Information*. In setting forth his main principles, Tufte notes, "graphical excellence is the well designed presentation of interesting data—a matter of substance of statistics, and of design." He further contends that graphics must "give to the viewer the greatest number of ideas in the shortest time with the least ink in the smallest space"[3] One of Tufte's main criticisms is that charts are often disproportional to actual differences in the data presented.

Examples of Distorted Graphics and Chartjunk

There are probably as many ways to distort graphics as there are graphic types. The Rogues' Gallery at the end of this chapter shows several examples of misrepresentation and describes the errors involved.

Tips for Avoiding Chartjunk and Deceptive Graphics

The guidelines given for making the various types of graphics will help you avoid making misleading or distorted visual aids. To summarize, follow these conventions for making data graphics:

Tables
- Line up numbers along the units column or the decimal point and use the same number of decimal points for each number.
- Arrange items in rows in some logical order.

---

[2]Daryl Moen, "Misinformation Graphics," *Aldus Magazine*, no. 2 (1990): 64.

[3]Edward R. Tufte, *The Visual Display of Quantitative Information* (Cheshire, CT: Graphics Press, 1983), 51.

Pie Charts
- Do not use 3D "hockey pucks" or angled views.

Bar/Column Charts
- Make all columns and bars the same width; the only variable should be length.
- Do not use log scales.
- Do not use 3D skyscrapers or cylinders.
- Use zero as your baseline for columns and bars.

Line/Alpine Charts
- There should be only one vertical y scale.
- It is acceptable to convert line charts to area charts, but don't combine charts in one visual.

All Data Graphics
- Compare apples to apples; don't mix charts.
- Limit background grids to the barest essentials; use thin grid lines and change them from black to gray.
- Indicate the sample size used when collecting data for statistics and the margin of error.

6.6   REPRESENTATIONAL GRAPHICS

Data graphics offer a way to show your audience how numbers or ideas compare. Representational graphics help your audience understand what something looks like, how it relates to various processes, or how it works. The two main types of representational graphics are those that use pictures—models, photography/video, and drawings—and those that use symbols, such as maps, flowcharts, and organization charts.

Pictorial Graphics: Models, Photos, and Drawings

Models

Pictures are nice, charts and diagrams help, clear verbal descriptions are very useful, but there is nothing that can compare to showing your audience the real thing. Bringing in a real example of what you are talking about is the most effective way of providing concrete evidence to an audience, combining vision, touch, taste, smell, and sound.

The next best thing may be to show them a model or mock-up. This is easy when the subject is, for example, a two-dimensional object like a website or a document. In that case, you can project it on a screen so the audience can see what it looks like. It is more complicated when the subject is three-dimensional, such as a new design for an appliance. Many designers today use computer-assisted design (CAD) programs to create 3D images. These can be used as presentation visuals. The newest versions of this software permit an object or a site to be experienced in four dimensions—they can show changes over time!

Models are often used in construction and architectural presentations, where they are combined with maps and artist's impressions to convey a sense of what a structure will look like. On big jobs, it is becoming more common to find presentations featuring computerized virtual reality tours and 3D flythroughs of projects. These animations are created by professionals and cost a lot to produce, but software prices continue to drop and the programs keep getting easier to use, so it is likely that this kind of graphic will become increasingly popular.

Here are some tips for using models as visual aids in presentations:

- Show the audience the "real thing" whenever possible. If possible, pass the object around, hold it up high so everyone can see it, or give everyone a sample. Make sure you label objects you pass around—by the time these get to the back row the audience members sitting there will have forgotten what they are looking at.
- Sometimes, when an object is dirty or dangerous, it is better to put it in a display case where it can be seen but do no harm. If you don't, Murphy's Law says someone will drop it, spill it, or get a nasty cut on it.
- Solve problems of small scale by placing the object on an overhead document camera and zooming in.
- Models and mock-ups must look professional or they will hurt your credibility. If you don't have model-building skills, you will have to pay a professional.
- If you really want your audience to be able to understand an object or a place, go beyond visual aids and find ways to reach their senses of sound, touch, taste, and smell.

Photographs and Video

Digital still and video cameras have become indispensable tools for presenters who want to show an audience what something or someone really looks like. Many businesspeople and engineers carry digital cameras with them wherever they go to record what they see and experience and some cell phones can take photos and send them electronically to others.

VISITING WEBSITES DURING A PRESENTATION

When you want to show your audience what a website or computer screen looks like, you can use a "screen shot" to show them the site or insert a hyperlink and go directly to the site. It is not hard to have your computer take a picture of whatever is on the monitor and then import that image into presentation visuals.[1] Figure 6–15 shows a screen shot of a website.

If you show your presentation visuals on a computer that has an open Internet connection, you may use a hyperlink to connect the website to your presentation. To do this with PowerPoint, select an object or text box on the screen, pull down the **Insert** menu and click on **Hyperlink . . . .** Type in the URL of the website to which you want to link. When your presentation visuals are shown as a **Slide Show,** you can click on the object you selected and the computer will open a Web browser and go to that website.

Many Internet browsers display information toolbars across the top of the screen that can be distracting. If you use a website as a visual, it will look better if you can eliminate these information bars.[2]

If reliability is important, use a screen shot and have a hyperlink prepared as a backup.

---

[1]Windows computers have a **Print Screen** key on their keyboards. When you press **Control** and this key, the computer takes a picture of your screen, which is saved on the Clipboard. To insert the image into a PowerPoint slide, create a new, blank slide and press **Control + v.** After the image is pasted into the slide, you can manipulate it like any other image. On an Apple computer, press **Shift + Command + 3** to take a picture of the whole screen or **Shift + Command + 4** to take a picture of part of a screen. On older Apples, images are saved as "Picture 1," "Picture 2," and so on, on the hard drive. On newer Apples, images are saved as .pdf files on the Desktop. (If you want to save the picture on the Clipboard, hold down the **Control** key while pressing **Command + Shift +** [3 or 4]).

[2]To hide the toolbars in Microsoft Internet Explorer, pull down the **View** menu and check **Collapse Toolbars** or uncheck the **Button, Address, Favorites, Status,** and **Explorer** bars. If you use Netscape Explorer, pull down the View menu and uncheck the items under **Show/Hide.** The appearance of most other browsers can be altered in much the same way.

The three main types of photographs used in presentation visuals are scanned images, digital photographs, and websites. When are photos a good choice for presentation visuals? Consider using photos (or video) whenever:

- You want to show your audience something that is too large to bring into the auditorium. It may be the only way a civilian audience could experience something like an aircraft carrier in action.
- The object or place you want to show them is too far away to reach. For example, in a presentation on the Arctic Wildlife Refuge, photographs and video are a lot more convenient than flying everyone to Alaska.

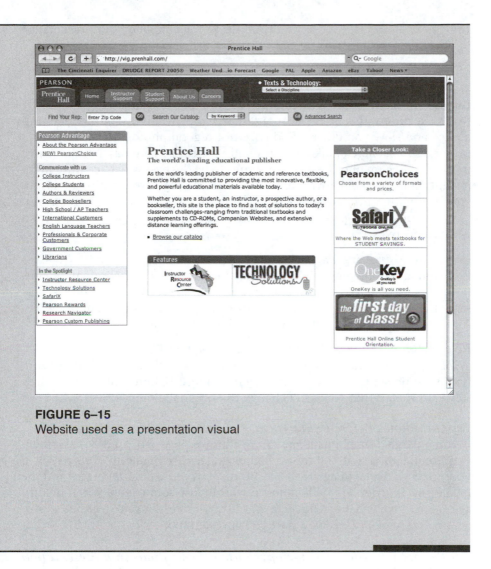

FIGURE 6–15
Website used as a presentation visual

- It is too dangerous for the audience to go where you want them to. For example, video is an excellent way to show how a blast furnace works in a description of the steel-making process.

While a model can show what a bridge will look like when completed, photographic images can help an audience understand what an object or person looks like in context—photos can show them the river and the banks the bridge will span.

Video adds another dimension to presentation graphics, because it can be used to show an audience the passage of time.

Here are some guidelines for using photos and video in presentations.

Choose High-contrast Pictures That Show a Lot of Detail
Don't use photos that aren't clear and sharp; they will only confuse your audience and slow you down. If you are not a good photographer, or if your camera isn't capable of taking high-quality pictures, you may have to hire a professional. Film photography continues to be the best way to capture high-resolution images, but advances in digital technology are catching up fast. If you start with a printed photograph, you will have to use a scanner to import pictures into a presentation. Some scanners can import images from 35mm negatives.

Crop Out Distracting Parts
Professional photographers understand the importance of eliminating clutter in a picture, so they crop out anything that distracts from the photo's message. If you use a photo for a presentation graphic, trim away anything that isn't essential before you show your audience the image.

Be Careful Not to Stretch Images So as to Distort Them
While it is a good idea to crop photos and to resize them so that they work with your presentation, be careful when you insert photos into presentation software visuals that you don't change the original dimensions.

Try Out Photos Before Using Them
If you enlarge a small image on 35mm film to the size of an 8′ × 12′ screen, you are magnifying it 100 times. Digital images are made up of tiny dots. If you project a digital image with an LED display, those dots get bigger and the image gets blurrier. You will almost certainly lose some detail when the image is enlarged by a projector. To avoid surprises, test out the computer, projector, and screen before you start your presentation.

Get Permission Before Using Copyrighted Illustrations in a Presentation
The rules against plagiarism require students to cite sources for any borrowed work, but they are not necessarily required to obtain permission prior to using it in a presentation. In most cases, the use of someone else's work by a student who is presenting in an educational setting and who is not paid for his or her work will fall under the "Fair Use" exception to the copyright law. As a general rule, though, do not use copyrighted photographs, video, or drawings to illustrate a presentation without getting permission.

Technical Drawings
Technical drawings are preferable to photographs when specific views are more important than photographic detail. These drawings are important tools for companies that produce or use technical products and can accom-

pany speeches that cover instructions, reports, and proposals, for example. Drawings that used to be produced by hand are now usually created by CAD systems. Like photographs, drawings can be scanned and converted to a digital format for use with presentation software.

PowerPoint has powerful drawing tools that make it easy to create objects and design elements. To use these tools, pull down the **View** menu, click on **Toolbars** and select **Drawing.** The **Drawing** toolbar contains a button that lets you select from among a dozen **AutoShapes.** The other buttons permit you to customize and combine shapes. While PowerPoint doesn't quite match a dedicated 3D Modeling or CAD program, it is extremely versatile.

Follow these guidelines for using technical drawings.

Limit the Amount of Detail
Keep drawings in oral presentations as simple as possible. Use only the level of detail that serves the purpose of your speech and satisfies your listener's needs. Figure 6–16 is an example of an exploded-view drawing that might be used as a visual in a speech about maintaining home heating systems. The purpose for using it is to help the audience focus on one component of the thermostat—the lever and attached roller. Completed on a CAD system, a speaker drew this picture so that the location of the arm can be seen easily.

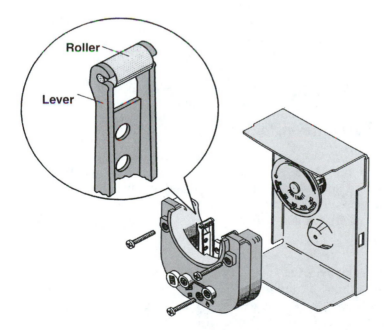

FIGURE 6–16
Technical drawing

Label Parts of Drawings

A common complaint about drawings used in speeches is that they include too much detail. Use callouts to label parts that are relevant to your speech. Make sure the callouts are readable from every seat in the audience. The simple labeling in Figure 6–16 fulfills these objectives. In complex drawings, avoid cluttering the illustration with too many labels.

Choose the Most Appropriate View

Drawn illustrations permit you to choose the level of detail needed. In addition, drawings offer you a number of options for perspective or view:

- **Exterior view.** Shows surface features with either a two- or three-dimensional appearance
- **Cross-sectional view.** Shows a "segment" of the object so that interiors can be viewed
- **Exploded view.** Shows relationship of parts to each other by "exploding" the mechanism—see Figure 6–16 for an example of how this looks
- **Cutaway view.** Shows the inner workings of an object by removing part of the exterior

Number and Title Each Drawing

Make it easy for your audience to understand what they are looking at as soon as the visual appears. Numbers and titles will also make it easier for you and your audience to refer to a particular visual and return to it if you are asked about it. It is a lot simpler to find a visual called "Fig. 5a New Flush Mechanism: exploded view" than to have to figure out which visual is which.

Symbolic Representations of Reality

Models, photos, and drawings are extremely useful for illustrating presentations, but there are times when it makes more sense to use a symbolic representation of reality, such as maps, flowcharts, and organization charts. For example, an aerial photograph can show what a highway looks like, but a map is so much easier to follow when you want to understand a plan for widening the highway.

Maps and Scale Drawings

Maps come in hundreds of different types, like political, geological, topographical, and cultural. They are excellent tools for studying and learning about a topic and modern cartographers, with the help of a geographic information system (GIS) work with enormous databases to extract information and create incredibly detailed maps.

Other map-like visuals include scale drawings, floorplans, and landscaping designs. Like maps, a well-drawn scale drawing can communicate an impressive amount of information. Maps of bodies of water are called *charts* and often have lines (isobars) and symbols indicating depths.

Maps and scale drawings can create problems when used as visual aids in presentations. Here are some ways of coping with these problems.

Eliminate Unnecessary Details

The biggest problem with most maps is that they have too much detail, which makes them harder to read quickly. If you must use a map, find the simplest one you can. If you have a photo manipulation program, such as Adobe Photoshop, you can remove excess information. You may want to start with an outline map and add the details you want your audience to see.

Always Show Scale

Maps, especially those showing construction sites, need to have a clear scale. Without a scale, it can be hard to determine how large things are and where they are in relation to one another. American audiences may be content with a scale given in English measurements (inches, feet, yards, miles, acres), but outside the United States you must include a metric scale.

Provide a Legend When Necessary

It is not always clear what map symbols mean. To make it easier for your audience to interpret symbols, include a legend with the map visual.

Always Indicate North

To help audience members orient themselves to the area shown on a map, add an arrow indicating which direction is north.

Flowcharts and Schedule Charts

Flowcharts tell a story about a process, usually stringing together a series of boxes or other shapes that represent separate activities. They have a reputation for being hard to read, so take extra care in designing them. Technical writers are often asked to write process descriptions; flowcharts are the tools used to illustrate those documents. Figure 6–17 shows a "top-down" flowchart that describes the process of using PowerPoint to create a top-down flowchart.

Using Flowcharts

In creating a flowchart, follow these guidelines:

- **Present only overviews.** Don't put details into a flowchart used as a presentation graphic. Your audience usually prefers a flowchart that gives only a capsule version of the process, not all the details. Reserve

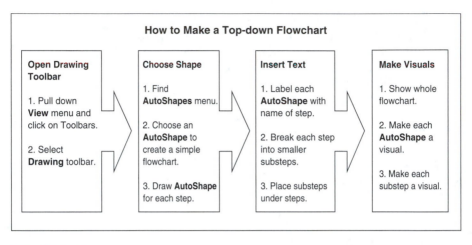

FIGURE 6–17
Top-down flowchart

your list of particulars for the text of the speech or put them into a handout.

- **Limit the number of shapes.** Flowcharts rely on shapes to relate a process—in effect, to tell a story. Different shapes represent different types of activities. This variety helps in describing a complex process, but it can also produce confusion. For the sake of clarity and simplicity, limit the number of different shapes in your flowcharts. Microsoft PowerPoint includes a set of conventional flowchart shapes (mostly related to computer data processing) in the **AutoShapes** menu that is part of the **Drawing** toolbar. These can be useful, but you can use almost any **AutoShape** as a text box and make it part of a flowchart.

- **Provide a legend when necessary.** Simple flowcharts rarely need a legend. The few shapes on the chart may already be labeled and the arrows make processes obvious. When charts get more complex, however, include a legend that identifies the meaning of each shape.

- **Run the sequence from top to bottom or from left to right.** Long flowcharts may cover an entire page. Yet they should always show some degree of uniformity by assuming either a basically vertical or horizontal direction. Use arrows to help your audience understand the direction of flow.

- **Label all shapes clearly.** In addition to a legend that defines meanings of different shapes, the chart usually includes a label for each individual shape or step.
  1. Place the label inside the shape.
  2. Place the label immediately outside the shape.

3. Put a number on each shape and place a legend for all numbers in another location.

Schedule Charts (Gantt and Milestone) Guidelines

Modern business practices require the creation of a written plan of action, which is frequently accompanied by an illustration showing the different parts of the project arranged along a time line. These schedule charts have many names, two of the most common being Gantt and Milestone. Project Managers use software, such as Milestones Professional or Microsoft Project, to build and manage Gantt charts when they do a Critical Path Analysis or Program Evaluation and Review Technique (PERT). The example shown in Figure 6–18 is a very simple Gantt chart.

The larger and more complex the project, the more complex the schedule chart will be. A good schedule chart does more than just highlight tasks and times—it goes into great detail. It is not unusual in the construction industry, for example, to produce schedule charts that cover an entire wall.

The PERT is a network model that helps planners allow for randomness in activity completion times. PERT was developed fifty years ago for the U.S. Navy to coordinate the work of thousands of contractors. It can reduce both the time and cost required to complete a project.

With a PERT chart like the one shown in Figure 6–19, each activity is defined by a "bubble" indicating the person or group working on it. The bubbles are connected by arrows to show the progress of the project.

Many technical presenters who try to include a schedule chart run into problems when reporting on proposals and feasibility studies. If you use

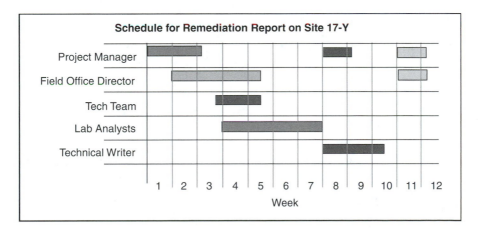

FIGURE 6–18
Simple Gantt chart

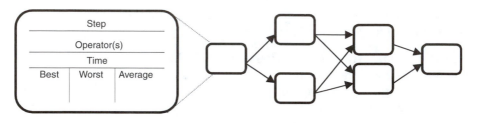

FIGURE 6–19
PERT chart

them as is (with all the details that make them useful), they won't be legible on a screen, and if you make them simple enough to fit on a screen, they are not very useful. One solution to this problem is to print the schedule chart and post it on the wall of the room where audience members can walk up and look at it during breaks. You might then put small portions of the schedule on visuals while you discuss them in your presentation.

Schedule charts usually include:

- The various parts of the project, in sequential order, listed along the vertical axis
- Time units shown on the horizontal axis
- Boxes containing information about each task and lines or separate markers showing the starting and ending times for each task

If you must use a schedule chart as a presentation visual, follow these guidelines:

- **Include only main activities.** Keep the audience focused on no more than ten or fifteen main activities. If more detail is needed, construct a series of schedule charts linked to a main "overview" chart.
- **List tasks in sequence from the top.** The convention is to list activities from the top to the bottom of the vertical axis. Thus, the observer's eyes move from the top left to the bottom right of the page, the most natural flow for many people.
- **Run all labels in a horizontal direction.** If audience members cannot easily read labels, they may lose interest. Try to avoid changing the direction of text.
- **Create new formats when needed.** Devise your own style of chart when it suits your purposes. Your goal is to find the simplest format for showing your audience, for example, when a product will be delivered or a service completed. You can buy project management software for

making schedule charts, but these programs are not designed for creating presentation graphics. Many people use Microsoft Excel to create schedule charts from spreadsheets. It may be easier to make a schedule chart from a table in Word or PowerPoint.

- **Be realistic about the schedule.** Schedule charts should never be thought of as unchangeable. In fact, you can almost count on having to make revisions as soon as you print one out. If your deadlines aren't feasible, they can come back to haunt you. As you set dates for activities, be realistic about the likely time something can be accomplished.

Most schedule charts show future events and deadlines, but you can also use them to show the past and indicate how well your organization has done so far in accomplishing its goals.

Visuals for Showing a Problem-solving Process

Business and industry today rely upon the written proposal as a way of adapting to changing circumstances. A proposal is a document that explains why an organization should start, alter, or stop doing some activity. For example, a county board interested in building a new sewage treatment plant might invite engineering firms to submit proposals for doing so. The written proposal is often followed by an oral presentation, so it has become increasingly important for technical experts to be able to explain to an audience how they reached a conclusion and why they are recommending a particular solution.

The format described here is one that can be used to illustrate the presentation of a proposal described in Chapter 5. You will recognize it if you are familiar with the format of a typical technical proposal. Each of the four steps could be turned into a visual, making it easy for your audience to understand how your proposal would work.

1. **Describe the problem.** Tell the audience how you know that there is a problem—What tests have been performed and data collected? What reasons are there to think it is significant and worth solving? Show a list of "symptoms" while you discuss them.
2. **Show cause and effect.** Explain what, in your professional opinion, is causing the problem and why you are eliminating other possible causes and explanations. List possible causes on a visual and discuss the list.
3. **Discuss what solutions are available.** Discuss what your experience has been and what your research into the problem has found. Explore the best alternatives and compare them to one another using relevant and valid criteria. Show comparisons on a grid with alternatives on one axis and criteria on the other.

4. **Recommend a solution.** Describe to your audience how your recommended solution would be implemented and describe your qualifications to implement the solution.

Organization Charts

Organization charts reveal the structure of an organization—people, positions, or work units. The challenge in producing this graphic for a large organization with many levels and many positions at each level is keeping it simple while trying to show enough detail. At the same time, you have to make certain the arrangement of information reflects the organization accurately. You can make a traditional organization chart with PowerPoint by clicking on **Object** . . . under the **Insert** menu and selecting **Microsoft Organization Chart.**

Use Linear Boxes to Emphasize Hierarchy

This traditional format uses rectangles connected by lines to represent some or all of the positions in an organization. Because high-level positions usually appear at the top of the chart, where the attention of most readers is focused, this design tends to emphasize upper management. Figure 6–20 shows a typical organization chart. The vast majority of organizations use charts with these boxes, so if you decide to use organization charts as a visual aide, this format is what most audiences will expect to see.

My Summer Job — Customer Service Rep for a Multinational Corporation

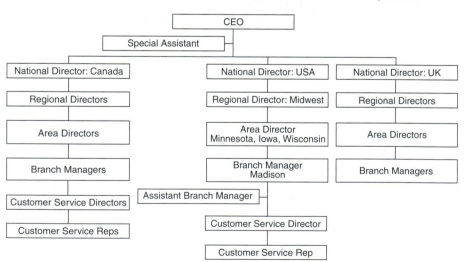

FIGURE 6–20
Organization chart

Connect Boxes with Lines
Solid lines show direct reporting relationships; dotted lines show indirect or staff relationships.

Try Other Designs to De-emphasize Hierarchy
One form of organization chart uses an arrangement of concentric circles to give more visibility to workers outside upper management. These are often the technical workers most deeply involved in the details of a project. The circular format is not common and may require extensive explanations.

Use Varied Shapes Carefully
Like flowcharts, organization charts can use different shapes to indicate different levels or types of jobs. However, beware of introducing more complexity than you need. Use more than one shape only if you are convinced this approach is needed to convey meaning to the audience.

Be Creative
When standard forms will not work, create new ones.

Computer Visualization and Plotting
Once upon a time, scientific testing was done with tools like a stopwatch, a thermometer, or a balance scale. The number of readings a researcher could collect was limited. Ways of displaying data were restricted to whatever could be plotted with a pen and graph paper. Computers have completely changed the way data is collected, and visualization software has created whole new ways of displaying data.

This software is still quite expensive and the learning curve for using these programs is still pretty steep. However, if you collect thousands of data points and want to build a numerical model to help your audience visualize what the data means, consider investing in one of these plotting software programs. To learn more about visualizing and plotting programs, try searching for "information visualization" on the Internet.

Rogues' Gallery of Problem Graphics

Law enforcement officials used to post photographs of known criminals on the walls of the station house where police officers would see them and be on the lookout for potential troublemakers. That was the origin of the Rogues' Gallery. In that tradition here are some problematic presentation graphics for your consideration.

Figure 6–21 shows a misleading bar chart with hash marks to indicate a break in the scale. The *y*-scale is misleading because the intervals are not equal. The hash marks through the tall columns are supposed to alert readers to the change in scale, but the may be overlooked very easily.

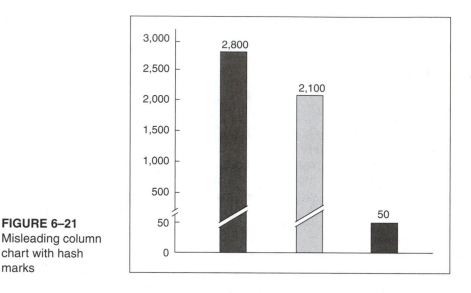

FIGURE 6–21
Misleading column
chart with hash
marks

Location of All Ding-Dong Convenience Stores

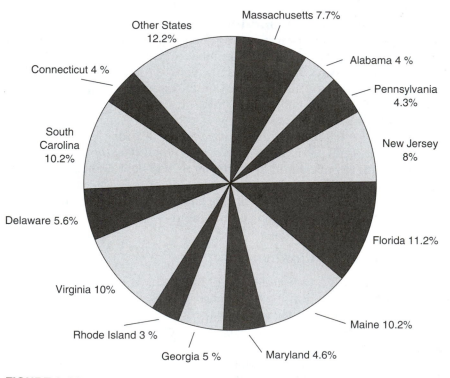

FIGURE 6–22
Confusing pie chart

The pie chart shown in Figure 6–22 has too many segments. It is impossible to determine what is significant. It violates the conventions of starting at 12 o'clock with the largest segment and continuing clockwise in decreasing order.

Figure 6–23 is based on a magazine ad that ran in the 1990s. At first glance, it seems like the advertised brand is significantly more reliable than the others. Once you realize that the y-axis starts at 93.5 percent, it becomes clear that there is only a tiny difference between the worst (A = 96.5 percent) and the best (Ours = 98.8 percent).

The "$25 bills" in the bar graph shown in Figure 6–24 are busy and distracting, but the real problem with the graph is that the bars are used to show area instead of length. The ratio of length from 1944 ($2.32) to 2004 ($25) is about 1:11, but the ratio of the area of the "$25 bills" for the same period is 1:105. That's misleading.

It's hard to get good information from the pie charts in Figure 6–25. You can see that more women than men were killed on the highway and by murderers, but there is no way of knowing what is significant without really studying both pies and comparing them segment by segment. Segments are labeled both inside and outside. The Transportation Incidents and Assaults and Violent Acts segments are made up of smaller segments; the only way you would know this would be if you realized that the same color is used for the subsegments.

The table in Figure 6–26 (created from responses to the 2000 Census) shows what languages Americans spoke at home in 2000 compared to the

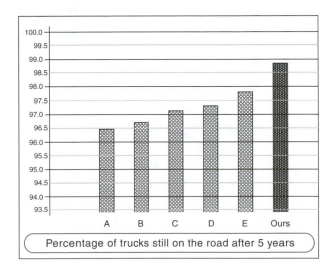

FIGURE 6–23

Misleading column graph (based on a magazine ad that ran in the 1990s)

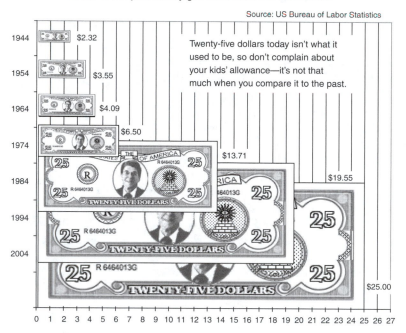

Where did all your money go? Inflated Dollars since WWII

Source: US Bureau of Labor Statistics

1944 — $2.32
1954 — $3.55
1964 — $4.09
1974 — $6.50
1984 — $13.71
1994 — $19.55
2004 — $25.00

Twenty-five dollars today isn't what it used to be, so don't complain about your kids' allowance—it's not that much when you compare it to the past.

0 1 2 3 4 5 6 7 8 9 10 11 12 13 14 15 16 17 18 19 20 21 22 23 24 25 26 27

FIGURE 6–24
Misleading bar graph

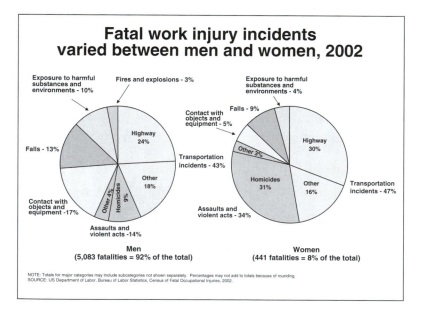

Fatal work injury incidents varied between men and women, 2002

Exposure to harmful substances and environments - 10%
Fires and explosions - 3%
Highway 24%
Falls - 13%
Transportation incidents - 43%
Other 18%
Contact with objects and equipment -17%
Other 4%
Homicides 9%
Assaults and violent acts -14%

Men
(5,083 fatalities = 92% of the total)

Exposure to harmful substances and environments - 4%
Falls - 9%
Contact with objects and equipment - 5%
Highway 30%
Other 3%
Homicides 31%
Other 16%
Transportation incidents - 47%
Assaults and violent acts - 34%

Women
(441 fatalities = 8% of the total)

NOTE: Totals for major categories may include subcategories not shown separately. Percentages may not add to totals because of rounding.
SOURCE: US Department of Labor, Bureau of Labor Statistics, Census of Fatal Occupational Injuries, 2002.

FIGURE 6–25
Confusing pair of pie charts

responses to the 1990 Census. It's confusing because the names of the languages spoken are jammed flush against the thick black borders to the left while the columns of numbers are flush right, leaving empty rivers of meaningless white. It would be much easier to understand if the person

Table 3. 20 Languages Most Frequently Spoken at Home for Population Age 5 and Older: 1990 and 2000

(Data based on sample. For information on confidentiality protection, sampling error, nonsampling error, and definitions, see www.census.gov/prod/cen2000/sf3.pdf)

	2000		1990	
Language spoken at home	Rank	Total Number of Speakers	Rank	Total Number of Speakers
United States	**(X)**	**262,375,152**	**(X)**	**230,445,777**
English only	(X)	215,423,557	(X)	198,600,798
Total non-English	(X)	46,951,595	(X)	31,844,979
Spanish	1	28,101,052	1	17,339,172
Chinese	2	2,022,143	5	1,249,213
French	3	1,643,838	2	1,702,176
German	4	1,382,613	3	1,547,099
Tagalog	5	1,224,241	6	843,251
Vietnamese[1]	6	1,009,627	9	507,069
Italian[1]	7	1,008,370	4	1,308,648
Korean	8	894,063	8	626,478
Russian	9	706,242	15	241,798
Polish	10	667,414	7	723,483
Arabic	11	614,582	13	355,150
Portuguese[2]	12	564,630	10	429,860
Japanese[2]	13	477,997	11	427,657
French Creole	14	453,368	19	187,658
Greek	15	365,436	12	388,260
Hindi[3]	16	317,057	14	331,484
Persian	17	312,085	18	201,865
Urdu[3]	18	262,900	(NA)	(NA)
Gujarathi	19	235,988	26	102,418
Armenian	20	202,708	20	149,694
All other languages	(X)	4,485,241	(X)	3,182,546

Note: The estimates in this table vary from actual values due to sampling errors. As a result, the number of speakers of some languages shown in the table may not be statistically different from the number of speakers of languages not shown in the table. In addition, the number of speakers of a language with a higher rank may not be statistically different from the number of speakers of a language with a lower rank.
1 In 2000, the number of Vietnamese speakers and Italian speakers were not statistically different from one another.
2 In 1990, the number of Portuguese speakers and Japanese speakers were not statistically different from one another.
3 In 1990, Hindi included those who spoke Urdu.
(X) - Not applicable
(NA) - Not available
Source: U.S. Census Bureau, Census 2000 Summary File 3.

FIGURE 6–26
Badly designed table

TAKING IT TO THE NEXT LEVEL

Oral presentations are considered so important in winning bidding competitions and in gaining investor support that companies are willing to spend thousands of dollars on visuals. 3D Modeling, CAD images, and Virtual Reality are rapidly becoming the norm when millions of dollars are on the line. The business world hires professional designers to ensure that their audiences will see only the most dazzling displays.

who designed this table had left some white space between the text and the borders and if the names of the languages were closer to the corresponding ranks. Interestingly, the designer decided to show the numbers of speakers and not the percentages, which would have been a lot more meaningful in making comparisons.

ENCOURAGEMENT

If you use a word-processor to write, you can create visuals on your computer. Keep them very simple and put the bulk of your time and energy into preparing what you want to say, and you will do fine.

EXERCISES

1. Find good examples of data graphics in a textbook, a newspaper, a popular magazine, and an academic journal. Do the images communicate information on their own or do you have to read a caption to understand them? Do they make understanding the topic easier and more efficient? How might you redesign these illustrations in documents for use in an oral presentation?

2. Find an example of a bad data graphic. Explain why your example is not helpful to viewers and what might be done to improve it.

7

Using Graphics

This chapter continues the ideas developed in Chapter 6, offering ideas on how to use graphics when giving a speech. It discusses the use of Microsoft PowerPoint and provides you with lists of things you should and should not do when using visuals. *Note:* Over the years, Microsoft has released dozens of different versions of PowerPoint. Your version may not have the features described here. If you run into problems, try the PowerPoint **Help** menu.

7.1   *SPECIAL ISSUES RELATED TO POWERPOINT*

About Microsoft PowerPoint

The first thing a lot of people think of when you mention visual aids is Microsoft PowerPoint. In just a few short years, this presentation software has taken on a dominant role in public speaking.[1]

Because it is such a powerful communication tool, PowerPoint has won many fans worldwide. However, some see serious flaws with the program and with the way people use and abuse it.[2] In fact, the debate over Power-Point has spread and intensified since the publication of *The Cognitive Style of PowerPoint* by Edward R. Tufte in 2003.[3] Tufte criticizes PowerPoint for encouraging oversimplification and poor graphic design. He argues that it has a negative impact on how people think about complicated topics and that its reliance on bullet lists severely restricts the flow of information.

---

[1]For a fascinating look at the origins of PowerPoint, see Ian Parker, "Absolute PowerPoint," *The New Yorker,* May 28, 2001.

[2]Criticisms of PowerPoint are not hard to find. Here are a few examples: Robert X. Cringely, "Now Hear This: If We're in Trouble, It's Probably Because People No Longer Really Listen," February 5, 2004, *http://www.pbs.org/cringely/pulpit/pulpit20040205.html* (accessed February 9, 2004); Rebecca Ganzel, "PowerPoint: what's wrong with It? Power Pointless," *Presentations,* February 2000, *http://www.presentations.com/search/article-display.jsp?vnu_content_id=1125436* (accessed April 26, 2002); Greg Jaffe, "What's Your Point, Lieutenant? Please, Just Cut to the Pie Charts," *Wall Street Journal,* April 26, 2000; and Julia Keller, "Is PowerPoint the Devil?", *Chicago Tribune,* January 22, 2003.

[3]Edward R. Tufte, *The Cognitive Style of PowerPoint* (Cheshire, CT: Graphics Press, 2003).

Whether or not you agree with Tufte and the other critics of PowerPoint, it would be foolish to disregard the norms of your organization's culture. In other words, it would be self-defeating to refuse to use PowerPoint if everyone else in your school, business, or other organization uses it to create visuals and if that is what management expects. What you can do, however, is learn to use the software effectively to illustrate your talk.[4] Let the visuals add to your presentation; don't let them replace your role as an active, engaged speaker.

What's Right with PowerPoint?

- It makes it very easy to combine text and pictures.
- Used carefully, it can make you look more professional and polished.
- All the bells and whistles are fun to use; you can add sound, movies, animation, links to websites, and so on.
- It has become the industry standard and you may be expected to use it.

What's *Wrong* with PowerPoint?

- Overuse of templates and clip art has made them cliché.
- Project Gallery, with its Auto Content feature, enables users with no understanding of basic presentation skills to create visuals, which kills creativity and spontaneity.
- People overload visuals (too much text and too many graphic elements) or strip down contents so much that they add nothing to the presentation.
- The graphing program in PowerPoint is weak and encourages presenters to make simplistic and hard-to-understand figures.
- The software is so easy to use that the temptation is to let the visuals drive the presentation, rather than the other way around.
- The time put into making fancy visuals would be better spent practicing and working on other, more important presentation skills.
- If you design your visuals on a computer other than the one you use for the actual presentation, the results are unpredictable: colors change, typefaces and graphics may disappear, and animations and other effects may not work.
- Visuals are not appropriate for all presentations. Audience expectations and the solemnity of the occasion should determine whether you use them or not.
- PowerPoint projections are too large for small audiences; use something more intimate.

---

[4]The website at *www.powerpointers.com* may help you understand how to use PowerPoint visuals. PowerPointers says their website is "dedicated to helping you communicate more effectively."

PowerPoint Guidelines

Microsoft's PowerPoint permits you to use your computer to create a wide variety of graphics, which then can be incorporated into a speech or written report. For a speech, you create a series of visuals called *slides*. These can be stored on a laptop or used on a different computer at the presentation site. In the room where you are giving your talk, you connect the computer to a video projector. While you are speaking, you move through the electronic slide show on the computer.

It's not complicated, but it can be confusing if you have not done it before. Here are a few basic guidelines for using Microsoft PowerPoint properly.

Use the Tutorial or Buy a Reference Book

Some versions of PowerPoint come with built-in tutorials that help you decide how, when, and where to use PowerPoint slides. If your version does not have a tutorial and you do not have an experienced PowerPoint user to assist you, consider investing in one of the many "how-to" books available. (Because there are many different versions of PowerPoint, make sure that the book you buy is written for the version you will be using.) You will discover that the software offers a dizzying array of charts, pictures, clip art, voice narration, and more. If you spend some front-end time getting to know the capabilities of the package, you will be more likely to make good choices later.

Start Simple and Stay That Way

With so many design options available, some PowerPoint users go overboard. They overdo the number of fonts, colors, shapes, screen movements, and other options. That may be a mistake. Stay with simple, clear principles of design so that your audience will focus on the material presented, not on the "bells and whistles" of the visuals. The KISS principle—"Keep It Short and Simple"—applies to PowerPoint presentations just as it does to speech text and other graphics.

Avoid Using Too Much Text

Studies of how people read periodicals have found that most of us prefer to look at pictures before we read text. It seems intuitive—looking at pictures is easy, reading text takes an effort. One of the most common problems with PowerPoint is that it encourages users to put a lot of text on the screen. Don't use PowerPoint as an excuse to present the kind of excessive verbiage you would never consider putting on more traditional graphics, such as overhead transparencies. Rather, learn to balance text and images to create a presentation that is visually interesting and easy to understand.

Use Keywords and Phrases—Not Sentences

The best way to use text on visuals is to make lists of keywords and phrases—
not sentences. Follow the guidelines in Chapter 6 for making and using lists.
If your slides include text, keep it brief and use them as lists of "talking
points" you will expand upon as you speak to the audience, not to the
screen. Reserve the use of whole sentences on visuals for direct quotations.
PowerPoint does not relieve you of a basic obligation of all speakers—to
maintain eye contact and to present material in an engaging fashion.

Check the Equipment Before the Presentation

More than one presenter has shown up with a PowerPoint-filled laptop, only
to find that the room does not include the right connecting cords for the
video projector. This and other technical headaches can be eliminated if you
or a trusted assistant check equipment ahead of time. It's also wise to carry
as much of your own equipment as possible.

Bring a Hard Copy of Slides

If something does go wrong with the hardware or software, you will be glad
you brought along a hard copy of the PowerPoint slides. They can quickly be
converted to overhead transparencies or handouts.

Don't Let the Technology Take Over

Your audience analysis will help determine the degree to which you should
rely on graphics. However, remember that most people still value the human
connection and prefer to hear a great speech reinforced by effective visuals,
rather than a high-tech visual display that eclipses a token speech.

7.2 INTERACT WITH YOUR VISUALS

No-nos When Using Visuals

Once you have decided to use visual aids to illustrate your presentation, you
need to be aware of the fact that many people find visuals annoying. In one
survey, these were the most common complaints about presentations accompanied by visuals:

- The speaker read the slides to us.
- The text was so small I couldn't read it.
- The visuals had full sentences instead of bullet points.
- The slides were hard to see because of poor color choice.
- Moving/flying text or graphics was distracting.

- The use of sounds was annoying.
- The diagrams or charts were overly complex.[5]

While some of these items are covered elsewhere in this book, here are a few things speakers should not do when using visuals:

Don't Talk to the Screen

It's acceptable to take a quick peek at the screen to make sure you have the right visual displayed, but you should face the audience the rest of the time. In addition to muffling your voice and making it hard to hear what you are saying, it is rude to turn your back on your audience. You can't make eye contact if you aren't facing forward.

Don't Read Graphics Word-for-word to the Audience

An especially irksome problem with PowerPoint is the tendency of some speakers to simply read text directly off the screen. This is often the result of the presenter having made visuals that are text-heavy. Reading to an audience may be perceived as arrogant or patronizing—unless you are addressing an extremely young audience of pre-readers, we all know how to read. It also makes it look like you are unprepared; you don't know the material well enough to look away from it.

Don't Gawk at Your Visuals

Don't give the audience the impression you have never seen your graphics before. It sometimes happens that a presenter is forced to step in at the last minute to give a talk prepared by someone else. But other than in that rare occasion, there is no excuse for acting as though you have never seen your visuals before. If you don't have the time to practice with your visuals, leave them home. Using them will only make you look like a rank amateur and hurt your credibility.

Don't Point Out Mistakes or Poorly Designed Graphics

Ethical presenters know that if you make a mistake that is material to your argument, you have an obligation to correct it. Small errors on your visuals or handouts may hurt your reputation for being careful and thorough (some people go nuts over spelling errors), but there is no reason to disparage your own work, thus hurting your credibility. If it's not important, leave it alone.

Guidelines for Using Presentation Graphics

Once you have prepared and practiced with your graphics, you're ready for the real thing. Following are some suggestions for handling graphics during the presentation itself.

---

[5]Dave Paradi, "Survey Shows How to Stop Annoying Audiences with Bad PowerPoint," *http:// www.communicateusingtechnology.com/articles/pptsurvey_article.htm* (accessed August 1, 2003).

Leave Graphics up for the Right Amount of Time

Because graphics reinforce text, they should be shown only while you address the related point in the text of the speech. For example, reveal a graph while saying, "As you can see from the graph, the projected revenue reaches a peak in 2005." Then pause and leave the graph up long enough for the audience to absorb your message. Remember these important points:

- A graphic outlives its usefulness when it remains in sight after you have moved on to another topic. Listeners will continue to study it and ignore what you are now saying.
- If you use a graphic once and plan to return to it, take it down after its first use and show it again later.

Use Handouts Wisely

Hard evidence—data, information, numbers, and experimental proofs—is what impresses technical experts. It will be part of any technical presentation where you want to persuade the audience and it is one way of establishing your credibility as a presenter. Nontechnical audience members, on the other hand, may not be able to follow, or be interested in following, a long, scientific or mathematical explanation. Therefore, it is your job as a presenter to find a way to get your technical audience the data they crave without boring the rest of your audience to tears. An excellent way to do this is by using handouts.

Rather than putting all the data into the body of your presentation, give out pages of calculations, graphs, and tables showing all your calculations. Your technical audience will be satisfied and your nontechnical audience will remain alert and awake. You can use handouts to provide information, to promote your business, or to advance your argument. Just don't make it look like it's all advertising—material that isn't relevant to the speech and appears to be promotional will likely be thrown away. Here are some more tips for making and using handouts:

- It's always a good idea to put your name and number on every page of the handout so people can call you if they have questions (or if they want to hire you to do another presentation).
- If you have several types of handouts and you want your audience to be able to follow along, consider making the different types on colored paper. It's a lot easier to ask people to "turn to the blue page" than to have them searching by heading or page number.
- Don't put your entire presentation on handouts. If you give the whole thing away, your audience will wonder why they had to attend. A much better choice is a bare-bones outline of what you will cover with lots of white space where audience members can take notes if they wish.

- If you want the audience to have the handouts before your presentation, have piles of copies set out near the door next to a "Please Take One" sign. If you want to pass them out during your speech, collate them and have them in neat piles you can easily access and hand out. Audience members may get impatient if they have to wait while you stop the speech to search for a bunch of papers.
- You want to make a strong, positive impression, so keep in mind that your handouts are a part of your speech that remains long after you are done talking. If you want to be remembered as a professional, your handouts should reflect your professionalism. Hire a graphic artist or take a course in document design if you want the best-looking documents. You don't have to be an artist yourself, but you should be professional if you want to be taken seriously.
- Print extra handouts. It's better than running short and worth spending the extra money.
- Use handouts to give your audience take-home readings, exercises, data printouts, and bibliographies.
- Depending on your topic, you may not even have to make your own handouts. Private industry, many government agencies, and educational institutions produce millions of documents that can be used as handouts. For example, if you are explaining how a particular product works, you may be able to get informational brochures from manufacturers at no cost. Similarly, many agencies of the Federal and state government put out free brochures and booklets that can be used as handouts.

Maintain Eye Contact

Making eye contact with your audience is an important delivery skill and it is even more important when you have graphics projected on a screen next to you. Don't stare at the screen showing your graphics while you speak. Maintain control of listeners' responses by looking at faces in the audience.

Make Arrangements in Advance

Arrange for projectors and screens ahead of time or plan on using your own equipment. Anything can happen when you use another person's audiovisual equipment: Whatever can go wrong, will—a new bulb burns out, there is no extra bulb in the equipment drawer, an extension cord is too short, the screen does not stay down, the client's computer doesn't read your disk. Many speakers have experienced these problems and the list is endless.

Even if the equipment works, it may not operate the way you are used to having it work. The only sure way to put the odds in your favor is to carry your own equipment and set it up in advance. However, most of us have to rely on someone else's equipment at least some of the time. Here are some suggestions:

- Find out exactly who will be responsible for providing equipment and contact that person in advance about details.
- Bring some easy-to-carry backup supplies with you—an extension cord, an overhead projector bulb, felt-tipped markers, and chalk, for example.
- Bring handout versions of your graphics to use as a backup.

In short, avoid putting yourself in the position of having to apologize. Plan well.

Hold Props Up High

Make sure people in all rows of the audience can see them and keep them up long enough so everyone has time to see them. If it is important to understanding your presentation, have the name of the prop written on a visual, whiteboard, or chalkboard. Small props should be passed around or enlarged by use of a document camera.

Get Out from Behind Computer Console

One disadvantage of using computer graphics is that, unless you have a wireless mouse or a helper to advance your visuals, you are stuck behind the console. Try to move around and gesture.

Blank the Screen When You Want the Audience to Look at You

Many projectors have a **Mute Screen** button that hides the image from the audience until you press it again. PowerPoint lets you press the **W** key on the keyboard to make the screen go white or **B** to have it go black—the image returns when you press the same key again.

Use a Pointer

Interact with your visuals by using a pointer deliberately to point to things on the screen. Office and teacher supply companies sell stick pointers. Laser pointers come in different colors and price ranges.[6] If you are projecting visuals from a computer, you may be able to use the cursor as a pointer or a pen. Here are a few tips:

- When using a pointer, find the object you want to point to on the screen, put the pointer on the object, turn around, and start speaking when facing the audience.
- Hold the pointer on the object long enough for everyone to see it.
- Don't wave a pointer around aimlessly.
- Don't bang a stick pointer on the screen.

---

[6]The longer, pencil-type laser pointers run on AA batteries, which are less expensive to replace. If you buy a laser pointer that uses an uncommon battery size, buy extra batteries and carry them with you.

- When pointing to a graphic with a stick pointer, use the hand closest to it. If you use the hand farthest away, you end up crossing your arm over your torso and, thus, turn your neck and head away from the audience.
- If your hands shake (from nervousness or from some other cause), the pointer will also shake, distracting any audience members who can see it. This problem gets worse if you are using a laser pointer. If you can't keep your hands steady, don't use a pointer.

7.3 *MASTERING POWERPOINT*

Although it is the most commonly used presentations graphics program, most users are self-taught. Few students have ever been shown these advanced features that can make a presentation more interesting. Microsoft has released many, many different versions of PowerPoint, so don't be surprised if the version you use doesn't have all the features described here.

PowerPoint Views: Slides, Notes, and Handouts

Microsoft PowerPoint comes with different views to help you while you are creating a presentation. Each of the views shows you different parts of the presentation. You may look at the five views individually by clicking on the buttons at the lower left of the PowerPoint window as shown in Figure 7–1.

If you choose the **Normal view** (shown in Figure 7–2), you will be able to see three of the other views at the same time in separate windows or panes: the **Outline pane,** the **Slide pane,** and the **Notes pane.** These panes let you work on all aspects of your presentation in one place. You can adjust the size of the different panes by dragging the pane borders.

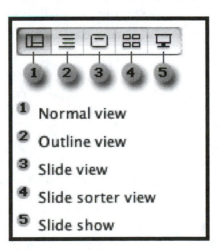

FIGURE 7–1
View buttons found at the lower left of a PowerPoint window (some Windows versions show only the Normal, Slide, and Slide Show views)

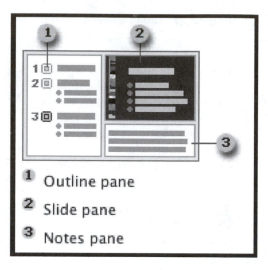

FIGURE 7–2
Normal view enables you to see the Outline, Slide, and Notes panes at the same time.

1 Outline pane
2 Slide pane
3 Notes pane

Outline Pane

You are supposed to be able to use the **Outline pane** (shown on the left in Figure 7–2) to organize and develop the content of your presentation. Unfortunately, everything you type into your **Outline** will appear on your visuals, whether or not you want the audience to see it.

If you want to make notes to yourself—about when to pause, catch your breath, ask if there are questions, or give a verbal example—put this information into the **Notes pane.** Remember to print out a set of notes before your presentation, because, when you use your computer to project slides (using the **Slide Show** feature of PowerPoint), you won't be able to see the **Notes pane**. To print a set of notes, click **Print** to open a Print dialog box and select **Notes.** On an Apple computer, after you click **Print** and open a Print dialog box, select the **Microsoft PowerPoint** option and change the **Print What:** window from the default (Slides) to **Notes**.

You can use **Outline pane** (or **Outline view**) to type all of the text you want to appear on your visuals; it gives you an easy way to rearrange bullet points, paragraphs, and slides. **Outline view** is the same as **Outline pane,** except that when you use **Outline view** you can't see the visuals or presenter's **Notes** as you work on the text of the outline. (You can get complete directions for using the **Outline pane** or **Outline view** if you work with PowerPoint by clicking on the **Help** menu and typing in **outline.**)

Slide Pane

In the **Slide pane,** you can see how text and graphics look on each slide. You can edit text, add graphics, movies, and sounds, create hyperlinks, and add animations to individual slides. **Slide View** is the same as **Slide pane,** except that you can't see the **Outline** or presenter's **Notes** when you are working in **Slide View.**

Notes Pane

The **Notes pane** lets you make notes for yourself as a presenter. You can use it to write down notes to yourself about when to pause, sip water, catch your breath, ask if there are questions, or give examples and other information to the audience. To be able to use the notes you make on the **Notes pane** during a presentation, you must print them out on paper and have them in front of you when you give your talk, because you won't be able to see them on the screen.

If you want to have your visuals print with your notes, add the notes in **Notes Page** view.

Unlike **Normal, Outline, Slide,** and **Slide Sorter,** there is no **Notes** button at the bottom of the PowerPoint window. In order to see and edit the **Notes Page,** you must pull down the **View** menu and click on **Notes Page.**

Slide Sorter View

In **Slide Sorter View,** you see miniatures ("thumbnails") of all the slides in your presentation at the same time. This makes it easy to add, delete, and move slides, add timings, and select animated transitions for moving from slide to slide. You can also preview animations on multiple slides by selecting the slides you want to preview and then clicking. However, you can't see your outline or your notes when you are in the **Slide Sorter View.**

Combine Text and Images

For a more interesting speech, use a combination of text and images on your visuals. Remember that a good picture can save you a thousand words. Import photographs and drawings to help illustrate your talk. The **Insert** menu shown in Figure 7–3 gives you many options for adding images to your presentation visuals.

Clip Art

PowerPoint comes with a large collection of clip art, but many of these images have been used so often that they are cliché—stale and boring like the example shown in Figure 7–4. If you must use clip art, select images that are all of the same artistic type. You can find more clip art online at Microsoft Office Clip Art and Media, *http://office.microsoft.com/clipart.*

There are two ways to insert clip art into a PowerPoint presentation visual. You can pull down the **Insert** menu and select **Picture ▶ Clip art. . .** or click **Insert Clip Art** on the **Drawing** toolbar. Use **Format Picture. . .** on the **Picture** toolbar to resize and reposition clip art. If the image overwhelms the text on a visual, try making it partially transparent, as shown in Figure 7–5, by moving the **Fill Transparency** slider bar shown in Figure 7–6.

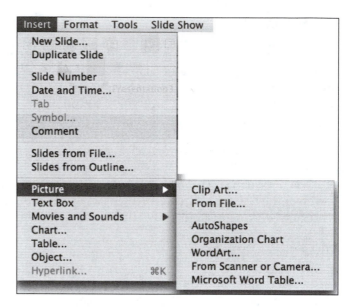

FIGURE 7–3
Insert Picture menu

FIGURE 7–4
PowerPoint clip art has been
overused and is now clichéd
and boring

The image on the left shows a piece of clip art as it appears with a **Fill** of 0% transparency. The image on the right is exactly the same, only with a **Fill** of 50% transparency, which makes it less distracting.

FIGURE 7–5
Make clip art slightly transparent and it will be less overwhelming (Windows users may have to ungroup the clip art before changing the transparency, depending on the version of PowerPoint being used and the format of the clip art)

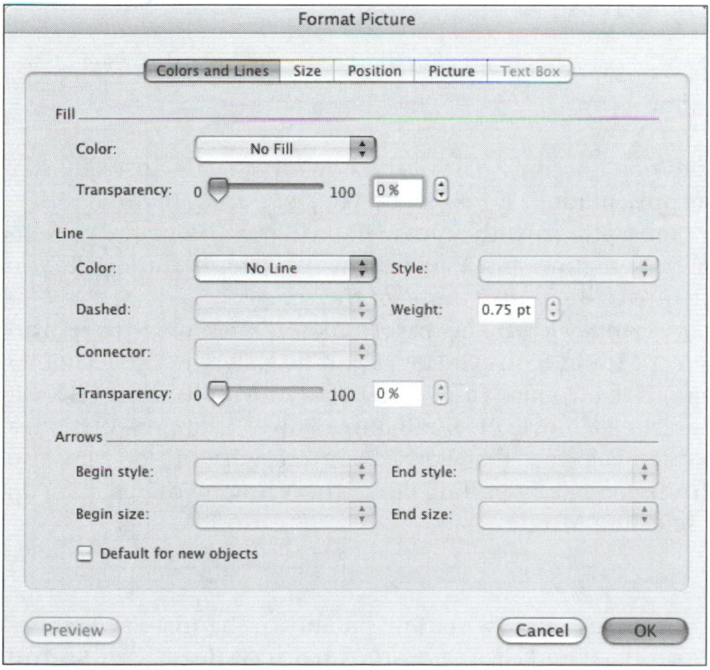

FIGURE 7–6
Format Picture menu

To insert an animated GIF[7] picture from the Clip Gallery, click on **Insert Clip Art** and then click **All Motion Clips** on the **Show** pop-up menu. To insert an animated GIF picture from a file, on the **Insert** menu, point to **Picture,** and then click **From File**. . . .

Photographs

To select a photograph or other image and place it into a presentation visual, pull down the **Insert** menu (see Figure 7–6) and click on **Picture** ▶ and **From file**. . . .

Follow the guidelines for using photos given in Chapter 6. Photos can also be made partially transparent. Just pull down the **Format** menu and select the **Picture**. . . toolbar under the **Colors and Lines** tab. Change transparency by moving the slider under **Fill.**

Inserting Movies, Audio Clips, Sounds, and Narration

Movies and Video Clips

PowerPoint lets you insert videos in the following formats:

- QuickTime (QT, MOV)
- QuickTime VR (QTVR)
- MPEG (MPG)
- Video for Windows (AVI)

Inserted movies don't become part of the PowerPoint file. Instead, they are linked to your presentation. If you show your presentation on a different computer, you must also copy the movie file when you copy the presentation or save the presentation as a PowerPoint Package. If you don't, you will just have a picture of the poster frame of the movie.

To insert a movie, pull down the **Insert** menu, click on **Movie and sounds** ▶ and select **Movie from Gallery**. . . (The gallery is a collection of animations and sounds that comes with PowerPoint) or **Movie from File**. . . . (For purposes of inserting a movie or sound into a PowerPoint presentation, a file can be anything in a format that PowerPoint recognizes.) You can also work with the **Movie** toolbar by pulling down the **View** menu, clicking on **Toolbars** ▶ and selecting **Movie.**

Sounds

To add sounds to a PowerPoint presentation, pull down the **Insert** menu select **Movie and Sounds** ▶ and click on **Sound from Gallery**. . . or **Sound**

---

[7]Graphics Interchange Format (GIF) is the standard bitmap format for Web graphics.

from File. . . .[8] The other two choices under Movies and sounds are **Play CD Audio Track**. . . and **Record Sound**. If you use sounds from a CD, the disk must be inserted into the CD drive of the computer you use when you give your presentation. To record sound, you will need to have a microphone connected to the computer.

Narration

If your computer has a microphone, you can record a voice narration for a PowerPoint presentation. Start by pulling down the **Slide Show** menu and click **Record Narration**. . . . Follow the directions on the screen to record your voice. *Note:* Some people find recorded sounds and narrations annoying. Microsoft recommends limiting the use of recorded narration to a few special situations—for Web-based presentations, for archiving a meeting so that presenters can review it later and hear comments made during the presentation, for individuals who can't attend a presentation, and for self-running slide shows.

Keyboard Controls for Screen, Pointer, and Drawing Pen

During a PowerPoint slide show, you can make changes to the cursor and the screen by using the right mouse button (or, on an Apple computer, by holding down the **Control** key while clicking the mouse button). Unfortunately, when you do this, the menu shows up on screen, so everyone can watch you fiddling with the controls. There is another way that is a bit more professional and just as easy—use the keyboard to control what appears on the screen (see Figure 7–7).

During a slide show, the arrow pointer will appear on the screen unless you decide you want to hide it or turn it into a pen for writing on your visuals. Figure 7–8 shows key board controls for changing the cursor. The arrow tool is a handy pointer for presenters with disabilities or those who must sit while presenting. The pen tool (sometimes referred to as a "Madden Pen"

---

[8]You can insert sounds into a PowerPoint slide show in the following formats:

- ◾ Audio Interchange File Format (AIF, AIFF, AIFC)
- ◾ Apple QuickTime Movie Sound (MOV, MOOV)
- ◾ Apple System Sound (SFIL)
- ◾ Apple System Resource Sound (RSRC, rsrc)
- ◾ CCITT A-Law (European Telephony) Audio Format (ALAW)
- ◾ CCITT U-Law (US Telephony) Audio Format (AU, SND, ULAW)
- ◾ Microsoft Windows Waveform (WAVE, WAV)
- ◾ MPEG Layer 3 Audio (MP3)
- ◾ Musical Instrument Digital Interface (MIDI, MID, KAR)

These keyboard keys will enable you to look like a pro:

- P = goes back to the <u>p</u>revious slide. This is the same as hitting the left arrow or up arrow keys, or the **Page Up** key.

- N = advances to the <u>n</u>ext slide. Exactly the same as clicking the mouse, hitting the right or the down arrow keys, or the **Page Down** key.

- B = <u>b</u>lack screen toggle. Exactly the same as hitting the period key.

- W = <u>w</u>hite screen toggle. Exactly the same as hitting the comma key.

- ESC (Escape) = Ends slide show.

FIGURE 7–7
Keyboard controls for PowerPoint visuals

after the football commentator who uses it frequently on TV) is extremely handy for emphasizing parts of a visual or for checking off items on a list. It takes a steady hand to use it during a presentation, so practice with the pen tool before you get up to speak. You can also change the color of the pen.

Caution: If you use a remote (cordless) mouse, put down the mouse when you are not using it. Otherwise, you might accidentally bump the mouse button and advance a slide before you are ready.

Customizing Backgrounds and Adding Logos

PowerPoint comes with a whole set of templates—background designs and color combinations—from which you can choose. Everyone who uses PowerPoint uses the same set of templates—so much so that they have become stale and boring. Fortunately, the program lets you create your own design using the Slide Master—a unique layout that becomes a template for background, text, and other images that you want to be the same on all your visuals.

Background

To give your presentation visuals a uniform look, all your slides should have the same basic background. Programs like PowerPoint let you customize the appearance of your slide background by changing its color, shade, pattern, or

The keyboard controls for making changes to the cursor are:

Apple computers:

- A = hide or show arrow pointer
- Command + P = change pointer to pen
- Command + A = change pointer to arrow
- E = erase pen drawing on the screen
- Control + mouse button = cursor menu appears on screen.

Windows computers:

- A = hide or show arrow pointer
- Control + P = change pointer to pen
- Control + A = change pointer to arrow
- E = erase pen drawing on the screen
- Right mouse button = cursor menu appears on screen.

FIGURE 7–8
Cursor controls

texture. When you change the background on the slide master, you can apply the change to all the slides in a presentation. Pull down the **Format** menu, select **Slide Background**, and click on **Fill Effects . . .** in the drop-down menu. Background choices for PowerPoint visuals are found in the **Fill Effects. . .** menu and include **Gradient, Texture, Pattern,** and **Picture.**

- **Gradients** are screens that gradually change from one color to another. They can be based on one color (shifting from dark to light), on two-color combinations, or on one of the preset combinations.
- **Textures** include images that are supposed to look like various types of "paper" and other materials. Many of them are unpleasantly distracting and should be avoided.
- **Patterns** permit you to combine two colors—one as background, one as foreground—in different ways to create patterns. Most of these are not suitable for slide backgrounds.
- **Pictures** includes most graphics files and it is also possible to use a photograph as a background for visuals. Be careful when doing this, because photos are often so cluttered and full of detail that it is hard to read text printed on top of them. (See Figure 7–9.)

Here is a technique for using a photo for a background that allows you to fade the image so that it doesn't interfere with the text:

1. Pull down the **View** menu, click on **Master ▶** and then select **Slide Master**.

2. Pull down the **Insert** menu to **Picture ▶** and then select **From File . . . .**

3. Find the photo you want to use for a background and click **Insert**. Do *not* click the **Link to File** box.

4. Once the picture has been inserted, you can make it brighter and increase the contrast so that it will work as a background.

FIGURE 7–9
Using a photo as a background

Text

PowerPoint visuals can be set up so the head, subhead, and text appear the same on each slide. For example, the default font for heads is Times 44, an adequate—but boring—typeface. To change the typeface, size, spacing, color, or bullet style for all the slides in a show, go to the Slide Master and make changes there. Make sure that the words stand out and that there is plenty of contrast between text and background. You can also put WordArt text on the Slide Master and it will appear on every slide.

Header and Footer

If you pull down the **View** menu and click on **Header and Footer. . .,** you will find that you can place a **Date and Time, Slide Number,** or **Footer** on every slide in a show. If you have any graphics or logos you want to appear on every slide, you can do so on the **Slide Master.** Some people find it very annoying having a corporate logo show up on every single visual, so try not to overdo it.

Linking to Websites: Hyperlinks and Action Buttons

Hyperlinks

If you want your audience to see what a website looks like, you can insert a picture of the screen (see Chapter 6) or add a hyperlink to your presentation. When you create a hyperlink, you turn some object—text, a shape, a table, a graph, or a picture—into a button. If you place the pointer on the button and click during a slideshow, PowerPoint will open a Web browser and the web-

---

BEWARE THE FICKLE INTERNET

Websites exist in cyberspace, an imaginary place where they can be easily changed or deleted without warning. One unverified story tells of a presenter who clicked on a hyperlink that was supposed to connect to a scientific website, only to have the screen filled with porn sites and pop-ups for sex toys. To avoid being embarrassed, *always* check your hyperlinks within 24 hours of presenting!

---

site will appear on the screen. Of course, this only works if you have an open Internet connection during your presentation. To get back to PowerPoint from the website, click **ESC** (escape). You can create a PowerPoint hyperlink by pulling down the **Insert** menu and clicking on **Hyperlink.**

Technology can be fickle and the one time you really need a hyperlink to work, it may not. As a backup, follow the steps described in Chapter 6 for taking a picture of your computer screen and pasting it into your visuals. If the website is too large to fit on a single screen, scroll down and take more screen pictures.

Action Buttons

PowerPoint comes with some ready-made action buttons (**Action Buttons** command on the **Slide Show** menu) that you can place on slides and define hyperlinks for. Use them when you want on-screen buttons for playing movies or sounds.

Drawing Toolbar

The **Drawing** toolbar has many features that will permit you to create simple graphics for your presentation. There are shadow and perspective effects you can use along with different fills to make various objects. If you can't see the **Drawing** toolbar on your computer screen, pull down the **View** menu and click on **Toolbars** ▶ and select **Drawing.**

WordArt

WordArt converts small amounts of text into objects you can manipulate on the screen. When you click on the **Insert WordArt** button, you will see a selection of styles. Choose one and you will see a panel called **Edit WordArt Text.** Type in the text you want to use and select a typeface. You can manipulate WordArt as though it were an object—change the size, shape, color of fill and line, make it partially transparent, rotate it, and more. WordArt has been discovered by young people and is extremely popular in grade school presentations. Therefore, adult PowerPoint users should probably avoid this tool.

AutoShapes

AutoShapes are geometric shapes that can be used as part of a drawing or as text boxes. They include the following categories:

- Basic shapes
- Block arrows
- Flowchart symbols
- Stars and banners
- Callouts
- Connectors
- Action buttons

Figure 7–10 shows how AutoShapes can be combined to create a simple drawing.

Text Boxes

If you have some text that is not part of your outline, but which adds to the information on a visual, you can place it in a separate text box. Pull down the **Insert** menu and click on **Text Box** or click on the **Text Box** button on the **Drawing** toolbar. When you click on the visual, the cursor will indicate that you are writing text in a rectangular box. Fill the text box with the words you want to appear on your visual.

FIGURE 7–10
Drawing with **AutoShapes.**
This old-fashioned TV Set
was made with **AutoShapes,**
a drawing tool that comes with
PowerPoint.

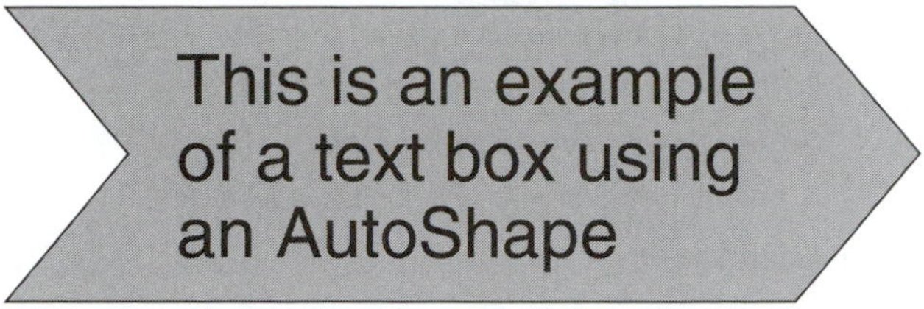

FIGURE 7–11
AutoShape as a text box

You can use most of the **AutoShapes** as borders for text boxes. After you have created your text box, click on the **Draw** button on the **Drawing** toolbar and select **Change AutoShape** ▶. You can turn your rectangular text box into any of the shapes on the menu. Figure 7–11 shows a box arrow used as a text box.

You can edit text in a text box the same you can edit any other text. Text in a text box will not show up in your outline.

If the borders of the box are too close to the text, you can change them by pulling down the **Format** menu and clicking on **AutoShape** . . . . Select the **Text Box** tab and change the internal margins so the border doesn't crowd the text. You can also adjust the way text wraps within the **Auto-Shape.**

Convert PowerPoint Files to Portable Data File (PDF) Files

If you like your typefaces and feel that you want to save all the hard work you put into designing your visuals, consider saving them in PDF format.

Portable Data File (PDF) is the format created by Adobe Acrobat. Images are viewable on all platforms, so it doesn't matter if you use Windows, Apple, or Unix systems. If you save your PowerPoint visuals file as a PDF, you can preserve your fonts, colors, and images exactly the way you created them (see Figure 7–12). *Note:* Some animations and artistic effects you can create in PowerPoint may not work properly in PDF format.

To run a screen show of PDF, you will need Adobe Acrobat or Adobe Acrobat Reader. Reader is a free, downloadable program any computer can run to display PDF files.

To use Acrobat or Acrobat Reader to run a slide show, start Adobe Acrobat or Adobe Reader, open the PDF file you want to use, pull down the **Window** menu and click on **Full Screen view.** Click the mouse key to advance slides. You can also use the arrow keys to go forward or back. To stop the show, press the **ESC** (escape) key.

Full Screen also offers a variety of slide transitions from which to choose. To change these, go to the **Preferences**. . . menu and select the **Full Screen** tab.

PowerPoint permits you to save files as PDFs by sending them to a virtual PDF printer.

1. When your presentation visuals are done, save the file first as a ppt. file.

2. Pull down the **File** menu and click **Print**.

3. Click on the **Save As PDF...** button.

4. Enter a name for the file in the **Save As** box.

5. Click **Save**.

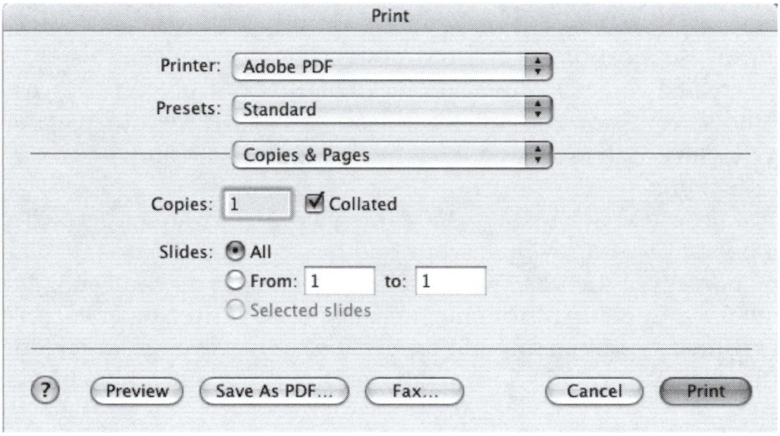

FIGURE 7–12
Some versions of PowerPoint permit you to save files as PDFs by sending them to a virtual PDF printer

Color Combinations for Visuals

If your goal is to design attractive, legible visuals, think in terms of high contrast between text or images and background. Some experts argue in favor of dark text on a light background, others favor light text on a dark background. It doesn't seem to matter very much as long as there is enough contrast to make the text readable.

We are used to reading black text on white paper, and that is a high contrast combination. But for projected images, if the background is too bright, then the light from the projector will illuminate part of the auditorium and the glare may make it hard for the speaker to see the audience. Here are some tips for selecting color combinations for visuals:

■ If you use light type on a dark background, pick a sans serif typeface and make it bold so it will stand out clearly (see Figure 7–13).

FIGURE 7–13
Sans serif type

This is a sans
serif typeface

- Although many of the PowerPoint templates use it, medium blue and light blue are poor color choices when you want to create a high-contrast image. If you use blue, go with navy or midnight for a dark background or very pale blue for a light background.
- Some degree of color blindness affects about 10 percent of the population. The most common form is an inability to distinguish red from green, so never put those two colors next to one another. Make it easier for color-blind people by using patterns and textures (as well as colors) to distinguish shapes.
- Colors will differ from computer to computer and from computer to projection system. Most color computer monitors come with a method of adjusting the way colors look on the screen. When you try to project an image from your computer on a screen, you may find that the colors don't look at all alike. If it is crucial to your presentation that the colors

COLORS AROUND THE WORLD

There is a whole research field devoted to studying the psychology of colors. Many colors have associations—some positive, some negative. For example, orange is associated with bargains and low prices while blue is associated with loyalty and solid thinking.

Be careful how you choose colors for your visuals when you present to non-American audiences. For example, white is the color of death in Japanese culture, but purple represents death in Brazil. Yellow is sacred to the Chinese but signifies sadness in Greece and jealousy in France. People from tropical countries seem to respond most favorably to warm colors; people from northern climates prefer cooler colors.

When in doubt, ask for help—you might feel silly calling the Korean Embassy to ask the cultural attaché if it's permissible to use a certain color combination in a presentation, but you would be far more embarrassed if you chose colors that offended people.

projected on the screen be true to those you used in your visuals, you will have to calibrate the projector. This can be time-consuming, even if you can find a manual that explains how to do it. If it is really important that the colors match, bring your own projector and adjust it ahead of time. Put in fresh light bulbs (old bulbs will have a different color) and always carry a spare bulb for your projector.

- Microsoft wants to help you. They have posted an interesting article called "Choose the Right Colors for Your PowerPoint 2002 Presentation" on the Internet at *http://office.microsoft.com/assistance/2002/articles/ppChoose RightColors.aspx*.

7.4 *FINAL THOUGHTS ON PRESENTATION VISUALS*

Three other points should be noted when discussing the use of PowerPoint.

Try Other Presentation Software

First, while Microsoft dominates the field, other presentation graphics programs exist. Harry Waldman, writing in the September 2002 issue of *Presentations* magazine, describes three alternatives to PowerPoint: Adobe Acrobat, Macromedia Flash MX, and Skunklabs Liquid Media.[9] Newer versions of these products and others are available. You will find more information about alternatives to PowerPoint in the Resources section of this book.

The Real Thing Is Still Best

Second, real samples and props that people can touch and hold are almost always better than 2D images on a screen. Whenever possible, let your audience see and touch the "real" thing and they will get a lot more out of the presentation. This is the best way to reach people who are kinesthetic learners (those who acquire information by holding and handling objects).

Hope for the Best, But Back up Everything

Third, to avoid problems, back up your visuals in a variety of media and be prepared to work without them if you have to.

---

[9]Harry Waldman, "Three Good Reasons to Stop Using PowerPoint," *Presentations,* September 2002, *http://www.presentations.com/presentations/technology/article_display.jsp?vnu_content_id=1731990* (accessed December 4, 2002).

TAKING IT TO THE NEXT LEVEL

A Dallas company, Teleportec, has introduced technology to transmit 3D holographic images of people over high-speed digital circuits. TeleSuite, based in Englewood, Ohio, makes and operates systems that use fiber optic broadband networks to link business conferences by a 14-wide, high-quality video screen that creates the impression that distant participants are in the same room with you.

ENCOURAGEMENT

Although it is not for everyone, new technology has made it much easier for presenters to illustrate their speeches. With a small investment of time, you can learn to make visual aids that will help your audience understand your topic. It takes time to master tools like computer software and video, so be patient. They say that a good picture is worth a thousand words, and this is especially true when it comes to using graphics with your presentation. The better you become at integrating visual aids into your talk, the easier it will be for you to present to diverse audiences.

EXERCISES

1. Find a chart, table, or other illustration from a periodical. Evaluate its effectiveness at communicating information. Analyze it in terms of audience and purpose. Discuss how it might be used in a presentation.

2. Observe an out-of-class presentation and pay attention to the use of visuals. Discuss what made them effective or ineffective. What might you do differently if you were giving the presentation?

8 Delivering Your Presentation

Have you ever heard a great storyteller speak? If so, do you remember how the spoken words communicated images and emotions and made you feel like you were really there as events unfolded? That's what delivery is all about. It is more than just the words that are spoken—it's the whole package of sights and sounds that you experience when observing a speaker.

Oral communication is extremely complex, which is why it may never be replaced by new technology. In addition to the words you choose to speak, there are a host of nonverbal signals you send to your audience. These may add to or detract from the clarity of your message. They may help or hurt your credibility. In fact, some researchers believe that nonverbal cues are more important to communicating than the words spoken. That's why it is so important to pay attention to nonverbal cues and to polish your delivery skills. There are many bad ways to deliver a talk; but there is no one right way to do it either. Learn by observing others, but remember that your delivery style is personal to you.

Delivery refers to all the things you do and say to communicate your message to your audience. In this chapter, you will learn about the different aspects of delivery and find some guidelines that will help you speak more effectively. You will learn about using your voice, adding humor to a presentation, polishing your delivery skills, and finalizing preparations for giving a great speech.

8.1 WORD CHOICE IS IMPORTANT

Choosing the Right Words

Here are some guidelines for choosing the right words for your presentation.

Use Neutral Terminology

Some people are easily offended by terminology that they don't consider politically correct. It may even distract them from hearing the rest of your message. Regardless of your own feelings about terminology, it is smart to avoid "loaded" terms that refer to race, sex, religion, or ethnicity that will needlessly offend audience members and keep them from attending to your message. It is especially important to avoid stereotyping—assuming that all the members of a group are alike—because many Americans find that offensive.

Let Your Audience Analysis Be Your Guide to Word Choice
The audience's experience and level of sophistication will determine what kind of vocabulary you use. Your terminology should be appropriate to your audience (which means it will be different for different audiences).

Choose the Correct Mode
You may have noticed that your mode of speaking is quite different when you are addressing a superior at work or school from the like, y'know, way you, like, talk to yer, y'know like, friends. Whatever.

People have different discourse modes for different situations. We use different vocabulary and terminology (slang/jargon) and vocal inflection with colleagues at work/school than with friends/peers, parents, or other professionals. Learn to switch modes when the occasion requires more formality. Some speaking modes entail vocal habits that are inappropriate in public speaking. It is important to use the proper mode for the situation or your credibility will suffer.

- **Start your presentation in a formal speaking mode.** Use complete sentences and speak in standard American English using correct grammar. You may later find that you have established rapport with your audience and can relax a bit. It is much easier to start at a formal level and become more informal later than to try to go in the other direction.
- **Avoid turning statements into questions.** Some speakers turn declarative sentences into questions by raising their voice tone at the end of the sentence. To hear how this sounds, try reading this sentence as though it had a question mark at the end(?). When you turn sentences into questions by raising your voice tone at the end, it sounds as though you are asking your audience a question instead of telling them what you know.

Stop Using Filler Words
Another problem habit related to informal speaking modes is the use of what speech therapists call "discourse particles"—filler words like *uh, OK,* and *uhmmm*—which can be very annoying to the audience. Speakers use filler words because silence makes them uncomfortable and because it is easier to start speaking when your vocal chords are already vibrating.

YOU NEVER KNOW WHAT WILL OFFEND PEOPLE

A few years ago I was asked to give a lecture to a large first-year class at the College of Engineering on how to prepare and deliver a presentation. Part of my talk was on the importance of having a strong introduction that really grabbed the audience's attention. I suggested that one should do whatever it took, even if it meant "starting a presentation with a circus parade, marching band, dancing girls, elephants, and acrobats."

A few days after my lecture, I received a call from a teaching assistant. Two young women in the audience had complained to him about my using "dancing girls" as an image in my talk. They felt that this terminology marginalized them.

I thought about it and then I apologized and promised that I would never, ever again use dancing girls as an example of how to get an audience's attention. I think this is a great example of how important it is to choose your words carefully.

Avoiding filler words presents a tremendous challenge to many speakers. When they think about what comes next or encounter a break in the speech, they fill the gap with words and phrases—*like, y'know, basically.* These gap-fillers are distracting, and once listeners hear a few, they start listening for more and stop attending to your presentation. It is hard to break these vocal habits without first being aware of when and how they occur. Here are some tips for getting rid of filler words:

- **Use pauses to your advantage.** Short gaps or pauses inform the listener that you are shifting from one point to another. In signaling a transition, a pause serves to draw attention to the point you make right after the pause. Note how listeners look at you when you pause. Do not fill these strategic pauses with filler words.
- **Practice with audiotape.** Tape is brutally honest: When you play it back, you will become instantly aware of fillers that occur more than once or twice. Keep a tally sheet of the fillers you use and their frequency. Your goal will be to reduce this frequency with every practice session. Once you have listened to yourself speaking on a recording, you will find it easier to stop yourself from, "uhhm, like," doing that.
- **Ask for help.** Present your speech to an individual who has been instructed to stop you after each filler. This technique gives you immediate reinforcement to stop using filler words.

Decide When to Give Definitions

Many speeches include terms that need to be defined for the audience. Nail down definitions. Practice definitions and explanations until they are as near-perfect as you can make them. Once you decide how you want them to sound, don't change them. Last-minute changes are sometimes obvious and will make you look indecisive and disorganized. Even though you may not need them, you should have prepared in your mind definitions for all the technical terms you use in your talk.

During your career, you may use technical terms known only to those in your profession. As a civil engineer, for example, you would know that a "tri-axial compression test" helps determine the strength of soil samples. As a computer professional, you would know the meaning of RAM (random-access memory), ROM (read-only memory), and LAN (local area network). When speaking to an audience that is unfamiliar with these fields, however, you need to define technical terms.

Good definitions can support findings, conclusions, and recommendations throughout your speech. They also keep audience members interested. Conversely, the most organized, well-written speech will fall on deaf ears if it includes terms that readers do not understand. For the sake of your listeners, then, you need to be asking questions like these about definitions:

- How often should you use them?
- Where should they be placed?
- What format should they take?
- How much information is enough and how much is too much?

Once you know definitions are needed in your speech, you must decide on their format and location. Again, consider your audience. Here are some guidelines for using definitions in a presentation.

Give Definitions Only When Necessary

Perhaps the most important question to ask is whether a definition is necessary, considering your audience, purpose, and time allotted for your speech. If your audience analysis reveals that most members of the audience are familiar with a word or expression, there is no need to define it. If the term is clearly outside the experience of most audience members and it is important to understanding your topic, you must have a clear concise definition somewhere in your speech. The toughest situation is when you are not sure if the term is familiar to your audience and it is not clear whether defining it will be seen as helpful or patronizing. For those cases, you should have a short, simple definition handy. This is called a "pocket definition" because you can "pull it out of your pocket" if you need it.

How do you know if you need to give a definition of a term to the audience? Here are some ways you can find out:

- Talk to the person who invited you to speak, and ask if audience members are familiar with technical terms.
- Pay attention to audience members' facial expressions and body language, and watch for signs of confusion or uncertainty.
- Stop momentarily and ask the audience members to raise their hands if they are familiar with a term. Some people are too embarrassed to admit when they don't know something. Take responsibility for the clarity of your work: Indicate that you are willing to go back and repeat what you just said if it wasn't clear. This relieves them of the burden of having to admit they didn't understand something.

Keep Definitions Simple

On rare occasions, the sole purpose of a speech is to define one or more terms. Most of the time, however, a definition just clarifies a term in a speech that has a larger purpose. Your definition should be as simple and unobtrusive as possible. Present only the level of detail needed by the listener. The more complicated the definition and the more definitions your audience needs, the more likely it is that you should consider creating a glossary of terms as a handout.

For example, in speaking to clients about your land survey of their industrial site, you might briefly define a transit as "the instrument used by land surveyors to measure horizontal and vertical angles." The main purpose of the speech is to present property lines and total acreage, not to give a lesson in surveying, so this short definition is adequate. But, if you need to explain metes and bounds, benchmarks, and other technical terms, create a glossary of terms and give it to your clients as a handout.

Choose from these three main formats (listed from least to most complex) in deciding the form and length of definitions:

- **Informal definition:** A word or brief phrase that gives only a synonym or other minimal information about the term
- **Formal definition:** A full sentence that distinguishes the term from other similar terms and that includes these three parts: the term itself, a class to which the term belongs, and distinguishing features of the term
- **Expanded definition:** A lengthy explanation that begins with a formal definition and is developed in depth

Use Informal Definitions for Simple Terms
Most Listeners Understand

Informal definitions appear immediately after the terms being defined, often as one-word synonyms. They give just enough information to keep the listener minimally informed about the topic. As such, they are best used with simple terms that can be defined adequately without much detail.

Here is a situation in which an informal definition would apply. Earlier in this chapter, the word *stereotyping* was used to describe something speakers should avoid doing. To make sure that readers understood what the word means, a short, ten-word definition followed (assuming that all the members of a group are alike).

For another example, assume that an engineer is presenting information to the owners of a piece of land where toxic waste has been found. The listeners do not need a fancy chemical explanation of *creosote*. They only need enough information to keep them from getting lost in the terminology, so the engineer might use an informal definition for the benefit of listeners:

> At the southwest corner of the site we found sixteen barrels of *creosote* (a coal tar derivative) buried under about 3 feet of sand.

Use Formal Definitions for More Complex Terms

A formal definition appears in the form of a sentence that lists (1) the *term* to be defined, (2) the *class* to which it belongs, and (3) the *features* that distinguish the term from others in the same class. Use it when your audience needs more background than an informal definition provides. Formal definitions define in two stages:

- First, they place the term into a *class* (group) of similar items.
- Second, they *list features* (characteristics) of the term that separate it from all others in that same class.

In the examples in Table 8–1, note that tangible terms (like *pumper*) and intangible terms (like *arrest*) can both be defined by first choosing a class and then selecting features that distinguish the term from others in the same class.

This list demonstrates three important points about formal definitions. First, the definition itself must not contain terms that are confusing to your readers. The definition of *triaxial compression test,* for example, assumes listeners will understand the term *shear failure* that is used to describe features. If this assumption were incorrect, then the term *shear failure* would need to be defined. Second, formal definitions may be so long that they create a major distraction in the speech. Third, the class must be narrow enough so that you will not have to list too many distinguishing features.

Put Expanded Definitions in a Handout

Sometimes a parenthetical phrase or formal sentence definition is not enough. If readers need more information, use an expanded definition with this three-part structure:

- **An overview at the beginning.** Include a formal sentence definition and a description of the ways you will expand the definition.

TABLE 8–1
Definition terms

Term	Class	Features
An *arrest* is . . .	restraint of persons . . .	that deprives them freedom of movement and binds them to the will and control of the arresting officer.
A *financial statement* is . . .	a historical report about a business . . .	that is prepared by an accountant to provide information useful in making economic decisions, particularly for owners and creditors.
A *triaxial compression test* is . . .	a soils lab test . . .	that determines the amount of force needed to cause a shear failure in a soil sample.
A *pumper* is . . .	a fire-fighting apparatus . . .	used to provide adequate pressure to propel streams of water toward a fire.

- **Supporting information in the middle.** Use listings on graphics as helpful format devices for the audience.
- **Brief closing remarks at the end.** Remind the reader of the definition's relevance to the whole speech.

Here are some ways to expand a definition, along with brief examples:

- **Background and/or history of term.** Expand the definition of *triaxial compression test* by giving a dictionary definition of *triaxial* and a brief history of the origin of the test.
- **Applications.** Expand the definition of *financial statement* to include a description of the use of such a statement by a company about to purchase a controlling interest in another company.
- **List of parts.** Expand the definition of *pumper* by listing the parts of the device, such as the compressor, the hose compartment, and the water tank.

- **Graphics.** Expand the description of the triaxial compression test with an illustration showing the laboratory test apparatus.
- **Comparison/contrast.** Expand the definition of a term like *management by objectives* (a technique for motivating and assessing the performance of employees) by pointing out similarities and differences between it and other management techniques.
- **Basic principle.** Expand the definition of *ohm* (a unit of electrical resistance equal to that of a conductor in which a current of one ampere is produced by a potential of 1 volt across its terminals) by explaining the principle of Ohm's Law (that for any circuit the electric current is directly proportional to the voltage and inversely proportional to the resistance).
- **Illustration.** Expand the definition of CAD/CAM (computer-aided design/computer-aided manufacturing techniques to automate the design and manufacture of products) by giving examples of how CAD/CAM is changing methods of manufacturing many items, from blue jeans to airplanes.

Obviously, long definitions might seem unwieldy within an oral presentation. For this reason, you need to think long and hard before you spend time on them. Most speakers prefer to give short definitions. Then they can provide a handout of definitions, for later reference by the audience.

Choose the Right Location for Your Definition

Short definitions are likely to be in the body of the speech; long ones are often relegated to technical handouts. However, length is not the main consideration. Think first about the *importance* of the definition to your audience. If you know that decision makers will need the definition, then place it in the speech—even if it is fairly lengthy. If the definition only provides supplementary information, then it can be eliminated or placed in a handout. Assuming that you have decided to define a term in your presentation, you have these choices for placing it:

- **An informal, parenthetical definition** can go in the same sentence as the term.

  Example: Tungsten filaments are strong and exceedingly *ductile* (which means that it is pliable, elastic, and can be drawn out into a wire or hammered into sheets).

- **In a separate sentence,** as a formal definition occurring right after a term is mentioned.

  Example: In the brain and the rest of the nervous system, serotonin acts as a *neurotransmitter*. Neurotransmitters are a class of brain chemicals that includes dopamine, norepinepherine, gamma-amino-butyric acid, and glutamate, and that can carry signals from one neuron to another.

- **Pocket definitions** can be pulled out and used at any point in a presentation. They are especially useful during a question-and-answer session, when your knowledge of what the audience did and did not understand often becomes clearer and you have a chance to correct errors, miscommunications, and omissions.

    Example: The questioner asked what do I mean by *sonoluminescence?* Sonoluminescence involves bombarding a liquid with sound waves to create tiny bubbles. The sound waves make the bubbles expand rapidly before collapsing so violently that they generate a flash of light and heat that is hot enough to trigger fusion in deuterium, a form of hydrogen.

If your speech is based on a written document, such as a proposal or an academic research paper, you can put a definition in a footnote at the bottom of the page on which the term is first mentioned, in a glossary, or in an appendix at the end of the document.

8.2 VOICE AS A NONVERBAL COMMUNICATION TOOL

Your voice is the most important tool you have for doing presentations. The quality of your speaking voice is one aspect of your presentation by which you will be judged.

There are men and women who earn a living with their speaking voices. Every radio commercial and most TV commercials features what is called 'voice talent'. Lawyers, politicians, teachers, and salespeople all use their voices to inform, persuade, and instruct. In our culture, having a strong, deep speaking voice adds to your credibility, while a thin, shaky, high-pitched voice can make you less believable.

A speaker's voice can enrapture the audience or make them want to flee the room. You can cultivate a voice that makes the audience keep listening. If you are not happy with the sound of your voice, there are several things you can do.

- A strong voice depends on your ability to push a lot of air out of your lungs, so the first thing to do is to focus on breathing properly.
- Work with a tape recorder and learn to control your breathing and your voice.
- Practice pronouncing hard words so that you can say them smoothly and authoritatively.
- Stand tall—poor posture can also make your voice sound weak.

If the problem is serious, you might want to work with a speech therapist or a voice coach.

Guidelines for Improving Your Voice

Here are some guidelines to help you capture your audience's attention with your voice.

Slow Down

New presenters sometimes stand up and start talking as fast as they can with no pauses in hope of ending the presentation as quickly as possible. When you rush, two things happen to your voice: You start slurring your words and you run out of air before you get to the end of the sentence. Slurring words makes it much harder for the audience to understand what you are saying. Running out of air means the last (and often most important) part of a sentence is less audible. When you address an audience, speak slowly enough so that they can follow what you say.

Learn to include pauses and silence between ideas. Many people are uncomfortable with silence and will rush to fill it with sounds. That's when filler words become a problem.

Pauses can be good for a presenter—they give you time to catch your breath, to check your outline, to observe the audience's reactions, and to sip some water. Don't be afraid to pause once in a while.

Speak Up or Use a Microphone

There is no point in giving a talk that the audience cannot hear. You must speak loudly enough so the entire audience can hear you. This takes energy and focus. If this is your problem, you need someone to coach you in how to project your voice or you can use a microphone.

Many factors affect your ability to project your voice. People from big cities tend to speak louder than rural residents. Men usually have louder voices than women because they have larger lung capacity and bigger larynxes (voice boxes). Trained singers and actors work on their voices and learn to project by exercising the muscles involved in speech: the vocal chords that vibrate and cause sound to come out, the diaphragm that pushes the air out of the lungs, and the muscles of the face and mouth that convert sounds into articulate words. Smoking and respiratory illnesses can limit a person's ability to speak loudly.

Warm Up

Learn to warm up your voice before speaking. No modern-day athlete would dream of working out or competing without warming up his or her muscles. The same principle applies to speaking. The discussion of microphone use described the voice as a product of three critical groups of muscles: the diaphragm, the vocal chords, and the muscles of the face and mouth that articulate sounds.

- **Control your diaphragm.** The diaphragm is a large muscle attached to the lower ribs, which separates the chest from the abdomen. Normally, it only works when you inhale and relaxes when you exhale. Deep breathing exercises will help you strengthen your diaphragm so you can project your voice farther.
- **Warm up your vocal chords.** Start slowly by humming, then try stretching by singing softly. When your voice no longer feels weak or sounds scratchy, move on to the next set of articulation exercises. Never scream or shout to warm up your voice. Don't smoke and, if you do smoke, quit.
- **Exercise your articulators.** The muscles of articulation—the jaw, tongue, lips, and soft palate—can be loosened with appropriate exercises. Try saying these sentences slowly and clearly. When you can do them at a conversational pace, speed up a bit.

Much of the flood comes under the hutch.
Sheep shears should be sharp.
Boots and shoes lose newness soon.
Peter Piper picked a peck of pickled peppers.
She sells seashells by the seashore.
The sixth sheep's sixth sheep's sick.
Ruth was rude to the youthful recruit.
Round and round the rugged rock the ragged rascal ran.
I slit the sheet, the sheet I slit, and on the slitted sheet I sit.

- **Build stamina.** Beginners are sometimes tempted to ignore the need for warming up their voices. Many speech classes require only short (seven to ten minutes long) presentations, so students see no real need for stamina. In business, industry, and academia it is not uncommon for speakers to talk for an hour or more. Regardless of what is required of you in school, you need to build up stamina to go beyond short in-class presentations. Make it a habit to use warm-up exercises before you speak and remember to use a microphone in front of big audiences.

Add Color to Your Voice

Good speakers know how to modulate their voices, so that their interest in and enthusiasm for their topic comes across. Avoid talking in a monotone (unless your purpose is to put your audience to sleep!). Learn to add color to your voice by practicing with a tape recorder. You may think you know how you sound to others, but the recording will enable you to identify your strengths and weaknesses. Here are some tips for making your voice more interesting.

- Don't ramble. Stay on track. Rambling is a sign of disorganization and a lack of practice.

WHEN TO USE A MICROPHONE (MIC)

Most of us can project our voices across a living room and, in an emergency we can find the strength to call across greater distances for help. In a lecture hall or auditorium, it takes more energy to project a voice so consider using a microphone. Here are some points to consider when deciding whether to use a microphone:

- Most people don't have trouble speaking loudly enough to reach an audience sitting three or four rows deep (about 50′ from the lectern). Beyond that distance, it is a good idea to use a microphone and amplification system.
- You should also use a microphone when speaking in a venue where noise may be a factor, such as during a plant tour or near construction or remodeling work.
- A microphone is almost mandatory when addressing an audience out-of-doors where you have to compete with street noises and other sounds.
- Always use a microphone if you are making an audio or videotape recording of your presentation. The built-in mics on camcorders are adequate for a lot of situations, but to get a high-fidelity recording of the audio portion of your talk, use a good microphone.

Here are some tips for how to use a microphone:

- Test it before you start using it. Adjust the volume with the help of someone who is willing to walk around the room and report on whether you are audible.
- Don't hold a microphone in front of a loudspeaker. That will result in annoying feedback.

- Avoid yawning. We usually associate yawning with feeling sleepy, but a lot of people also yawn when they are nervous or scared.
- Pause deliberately and try to emphasize the most important words in each sentence.

Use Cough Drops and Lozenges

If your throat is sore or your mouth dries out, speaking can be painful. It is a good idea to have a few throat lozenges or cough drops with you when you go to do a presentation. You can suck on them right up until the minute you are introduced. At that point, swallow them or dispose of them some other way; never give a speech with anything in your mouth.

- Don't tap the mic to see if it is working. Say a few words at your normal speaking volume to test it.
- Most mics have an on/off switch: some have two. Make sure they are on. Have an extra battery on hand in case you need one.
- If you have never used a mic before, practice with one to get used to it.
- Don't swing a microphone by the cord.
- Don't cover your mouth with the mic. When speaking, hold the microphone an inch or so in front of and an inch or so below your mouth. Your face and mouth are nonverbal communicators and it will be harder for your audience to understand you if you cover them up with the microphone. *Note:* When you see a performer concealing his or her mouth behind a microphone it usually means that you are hearing a pre-recorded song rather than a live performance. Your lips should not touch the mic.
- A handheld mic will start to become uncomfortably heavy if you are going to be speaking for more than ten minutes or so. Ask about getting a wireless microphone that leaves your hands free to gesture. A *lavalier* or *tie-tack mic* clips on to your clothing, whereas a small condenser *headworn* microphone can be hidden next to your cheek. If you use a wireless mic, make sure you turn it off whenever you leave the podium. There are some funny (mostly) anecdotes about what happened to speakers who left their mics on when they went to the restroom or out for a cigarette break.

It is especially important to remember to repeat back audience questions when you use a microphone, because it is more likely that other audience members will not be able to hear questions.

Get Help for Speech Problems and Accents

Students and businesspeople from other parts of the world often worry about whether American audience members will understand their accented English. The audience can adjust to your way of speaking if you make an effort to speak slowly and clearly and give them a chance to hear you speak a few sentences before you get into the content of the presentation.

Teenagers are notorious for mumbling and for turning declarative statements into questions by raising the pitch of their voices at the ends of sentences. This makes them sound less sure of themselves and hurts their credibility. Pay attention to the way you speak and you can break this habit.

If you have a serious speech problem, you might want to seek help from a speech or language pathologist.[1] A few sessions with a professional speech therapist can solve a lot of communication problems.

Speak With Energy and Enthusiasm

The real key to being a successful presenter is to speak with energy and enthusiasm. These two things can compensate for a truckload of problems. If you say it like you believe it, if you speak with passion, if you can convey your excitement to your audience, you can sell your audience on just about anything!

Use Both Your Voice and Your Body to Communicate

Nonverbal cues make up more than half of a presenter's message. Keep in mind that your facial expression, body language, and gestures all help audiences understand your speech.

8.3   *HUMOR: ADDING A LIGHTER TOUCH*

The discussion of organization noted that telling jokes is usually not a good thing for a speaker to do, because so many things can go wrong. There are, however, many other ways to inject a little bit of humor into your talk besides telling jokes. For example, you might use a humorous anecdote that makes a point. Or you might make a little play on words. Or you could show a cartoon (such as the one shown in Figure 8–1) that pokes fun of the way some professionals talk.

---

[1]Professionals in this field are accredited by the American Speech-Language-Hearing Association (ASHA) and receive a Certificate of Clinical Competence (CCC). You can learn more by looking under "Speech & Language Pathologists" in the Yellow Pages of the phone book or by going to the ASHA website at *http://www.asha.org.*

DILBERT® by Scott Adams

FIGURE 8–1
Cartoon   (DILBERT: © Scott Adams/Dist. By United Feature Syndicate, Inc.)

Here are some guidelines for using humor in a presentation. Just be aware that whenever you choose to use humor, you run the risk of offending audience members.

Appropriate Use of Humor

Here are some situations in which you might add a touch of humor to a presentation:

- To help make a point
- To make a point relevant and memorable.
- To make a serious topic a bit lighter
- To show the audience that you have a sense of humor and don't take everything seriously all the time

Not all presentation topics lend themselves to humor. For example, think carefully about your audience and your purpose before deciding to add something humorous to a talk on colon cancer or child abuse.

Many times it is better to make yourself butt of the joke rather than to target another person or institution. People will feel more comfortable if they are laughing with you at yourself.

When to Avoid Humor

There are a few kinds of humor that don't seem to work well in presentations. These are the ones to avoid:

- **Sarcasm.** People don't get it and will spend the rest of the presentation trying to figure out if you were being serious or not. Ironically, the same doesn't hold true for irony.[2]
- **Obscure allusions.** Mentioning people or events with which the audience is not familiar will result in blank stares instead of laughter. If you stop and explain them, the audience will feel that you are wasting their time or talking down to them.
- **Puns.** Some people really hate them. Perhaps the notion that a word can have more than one meaning makes them uncomfortable.
- **Dirty jokes.** Courts have held that telling dirty jokes contributes to a "hostile environment" in which harassment occurs. Use obscenities at your peril. Be prepared to defend yourself against accusations of sexual harassment.

---

[2]*Irony* comes from using words in a humorous way to mean the exact opposite of what they usually mean or from something happening that is the exact opposite of what one might expect. One example is the old story about the student who plagiarizes a paper only to discover that the original author is the very professor to whom the student submitted the paper. *Sarcasm* is making ironic comments to mock someone or to make fun of them. Calling someone who has just made a foolish mistake "Genius" is sarcasm.

Tell Anecdotes, Not Jokes

An anecdote is a story, related to your topic, which makes a point. It can be funny, ironic, or sad. It doesn't have a punch line, but it may have a moral or teach a lesson about life.

Avoid Humor with Hypersensitive Audiences

Use humor carefully if you want to avoid offending audience members. The earlier discussion of political correctness also applies to the use of humor. Something you consider hysterically funny could insult and offend sensitive audience members. Be extremely careful when discussing controversial issues. Avoid language, remarks, and "humor" that may hurt your credibility.

8.4   *GUIDELINES FOR DELIVERY*

You have prepared the speech and practiced it often. If you get nervous before such occasions, you are employing strategies for reducing anxiety. Now it's "showtime."

Although you want to appear natural to the audience, you do not want to display a casual manner. To keep people interested in the speech you must use techniques that are not evident in normal, informal conversation. Here are some guidelines for delivering the speech and for handling questions from the audience.

ANECDOTES

If you're not familiar with this word, it is from the Greek:

an- = not
ec- = out
dote = given

Thus an *anecdote* is the unpublished details, or the juiciest parts of the story. It is often confused with the word *antidote*, which is also from the Greek:

anti- = against
dote = given

An *antidote* is medicine given to counteract the effects of some toxic substance. The *antidote* to a boring speech is a poignant *anecdote*.

With practice, you will discover and refine the delivery techniques that work best for you. Follow the guidelines in this section to make each speech "showtime" without making it "showy."

1. **Speak in a clear, confident, loud voice.**

2. **Practice nonverbal communicators.** When you practice your presentation, you should also work on pronunciation, transitions, and gestures. When in doubt about pronunciation, ask someone who knows to help you. To establish your credibility as an expert on your topic, practice pronouncing technical terms clearly and correctly, without hesitation. At the same time, practice transitions from one part of your speech to the next and rehearse your gestures.

3. **Don't rush.** Speak at a steady pace.

4. **Keep sentences short, simple, and declarative.**

5. **Show enthusiasm for—and knowledge of—the subject.** Few, if any, listeners ever complain about a speech being too enthusiastic or a speaker being too energetic. But many, many people complain about dull speakers who fail to show that they themselves are excited about the topic. You may wonder, "How much enthusiasm is enough?" The best way to answer this question is to hear or watch yourself on tape. Your delivery should incorporate just enough enthusiasm so that it sounds and looks a bit unnatural to you. Remember—every presentation is, in a sense, "showtime."

6. **Speak vigorously and deliberately.** *Vigorously* means with enthusiasm; *deliberately* means with care, attention, and appropriate emphasis on words and phrases. The importance of this guideline becomes clear when you think back to how you felt during the last speech you heard. At the very least, you expected the speaker to show interest in the subject and to demonstrate enthusiasm. Good information is not enough. You need to arouse the interest of the listeners. Energy and enthusiasm enhance your credibility.

7. **Use appropriate gestures and posture.** Adopt appropriate posture and use gestures that will reinforce what you are saying.

   The Good. Good speakers are much more than "talking heads" behind a lectern. Instead, they do the following:
   - Use their hands and fingers to emphasize major points
   - Stand straight, without leaning on or gripping the lectern
   - Point toward visuals on screens or charts, without losing eye contact with the audience

   The Bad. The audience will judge you by what you say and by what they see, which makes videotaping a crucial part of your preparation. By working on this aspect of your speech delivery, you will avoid problems like these:

- Keeping your hands in your pockets; this is considered rude (the taboo against talking with your hands in your pockets originated long ago when people were worried that you might pull out a weapon).
- Rustling pocket change; remove it and keys from pockets beforehand
- Tapping a pencil, your note cards, or computer mouse.
- Scratching yourself. (As a rule, never ever touch any part of your body while doing a presentation that you wouldn't be comfortable touching in front of your grandma on nationwide TV.)
- Slouching over the lectern.
- Shifting from foot to foot. Move deliberately when you are presenting; if you decide to walk across the stage, do it without hesitation.
- Pointing your forefinger at audience members; this is considered very rude (this taboo comes from the idea that shaking the pointing finger at someone is associated with scolding and blaming). It is acceptable to point at people with two fingers or with an open palm.

The Ugly. Some hand gestures that are okay in the United States are considered obscene in other countries. Presenters should be sensitive to audience members' cultural backgrounds and find out which gestures are offensive. The following are acceptable in the United States but can be obscene in other parts of the world:

- Thumb touching forefinger to indicate OK
- Thumbs up
- V for victory sign

Finally, if you enumerate your main points by counting on your fingers, make sure you don't finish on you middle finger—that gesture is considered obscene just about everywhere.

8. **Move deliberately: Come out from behind lectern.** Lecterns are barriers between speakers and audiences because they conceal body language. Step out from behind the lectern on occasion to decrease the distance between you and your audience and to give them a chance to see your whole body. This will add to your credibility and create greater rapport between you and your audience.

9. **Scan audience and make eye contact.** Your main goal—always—is to keep listeners interested in what you are saying. This goal requires that you maintain control, using whatever techniques possible to direct the attention of the audience. The important things is that you look at the audience and that you don't stare at the ceiling or study your shoes. Frequent eye contact is one good strategy for persuading the audience that you want to communicate with them.

Some people find it hard to look others in the eye, but in our culture making eye contact is considered a sign of openness and honesty.[3] If you have trouble making eye contact, you will have to work on this. Learn to scan your audience, looking at each face for a second or two—longer than that is considered staring, which is rude. Don't try to avoid making eye contact by looking over their heads; this is obvious and they will assume you are watching the clock.

Listeners pay closer attention to what you are saying when you look at them. Think of how you react when a speaker makes eye contact with you. If you are like most people, you feel as if the speaker is speaking to you personally—even if there are 100 other people in the audience. Also, you tend to feel more obligated to listen when you know that the speaker's eyes will be meeting yours throughout the presentation.

Here are some ways you can make eye contact a natural part of your own strategy for effective oral presentations:

- **With small audiences:** Make regular eye contact with everyone in the room. Be particularly careful not to ignore members of the audience who are seated to your far right and far left. Many speakers tend to focus on the listeners sitting in the center of the audience. Instead, make wide sweeps so that listeners in all parts of the audience (shown in Figure 8–2) get equal attention.
- **With large audiences:** There may be too many people or the room may be too large for you to make individual eye contact with all listeners. In this case, talk to the whole room, but focus on just a few

---

[3]Eye contact is not acceptable in all cultures. In many parts of Asia, it is considered extremely rude to make eye contact with someone who is older or who has higher social status. To learn more about giving talks in other parts of the world, see *Kiss, Bow, or Shake Hands* by Terri Morrison, Wayne A. Conaway, and George A. Borden (Aron, MA: Bob Adams, Inc., 1994).

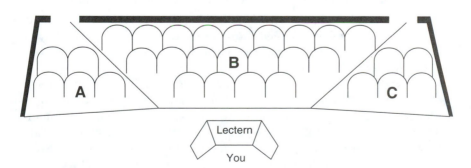

FIGURE 8–2
Audience sections

people in different parts of the audience. If possible, come down from the podium and walk around in front of the audience. This approach gives the appearance that you are making eye contact with the entire audience.

- **With any size audience:** Talk to all parts of the audience and don't worry about making eye contact. Occasionally look away from the audience—either to your notes or toward a part of the room where there are no faces looking back. In this way, you avoid the appearance of staring. Also, these breaks give you the chance to collect your thoughts or check your notes.

- **In front of a camera:** Follow the guidelines for where to look and where to talk given in Chapter 10.

10. **Converse, Don't Lecture.** Speakers today need to find ways of interacting with the audience and conversing, rather than lecturing. Here are some ideas for making your presentation more interactive:

- **Ask questions.** Find out what the audience thinks or knows about an issue. This is a great way to discover any misconceptions they may have and to correct them.

- **Collect information.** Ask your audience to share their knowledge and experience with you and with one another.

- **Have audience members write something and ask them to read it aloud.** Give audience members time to make a list or write a description and then have volunteers read them to the audience. This is similar to the previous item, but writing takes more time and effort, so you can expect more. Audience members can take their writings home as a reminder of the presentation (or you may collect these as a form of feedback on your talk).

- **Play a game or give them a puzzle to solve.** Games and puzzles are good ways to help audience members tap into their creativity. This can be valuable when the focus of your talk is about how to solve a problem.

11. **Use Rhetorical Questions and "What If" Scenarios.** Enthusiasm, of course, is the best delivery technique for capturing the attention of the audience. Another technique is the use of rhetorical questions at pivotal points in your presentation.

Rhetorical questions are those you ask to get listeners thinking about a topic, not those you would expect them to answer out loud. They prod listeners to think about your point and set up an expectation that important information will follow. Also, they break the monotony of standard declarative sentence patterns. For example, here is a rhetorical question used by a computer salesperson in proposing a purchase by a firm:

> I've discussed the three main advantages that a centralized word processing center would provide your office staff. But you may be won-

dering, "Is this an approach we can afford at this point in the company's growth?"

The speaker would follow this rhetorical question with remarks supporting the position that the system is affordable.

"What if" scenarios are another form of rhetorical question. They gain listeners' attention by having them envision a situation that might occur. For example, a safety engineer could use the following rhetorical question in discussing asbestos-removal services in a presentation to bankers:

> What if you repossessed a building that contained dangerous levels of asbestos? Do you think your bank would then be liable for removing all of the asbestos?

Rhetorical questions do not come naturally. You must make a conscious effort to insert them at points when it is most important to gain or redirect the attention of the audience. Be careful not to overdo the use of rhetorical questions—too many of these thought-provoking questions and the audience will lose track of your presentation. Three particularly effective uses of rhetorical questions follow:

- **As a hook at the beginning of a speech:** "Have you ever wondered how you might improve the productivity of your word processing staff?"
- **As a transition between major points:** "We've seen that centralized word processing can improve the speed of report production, but will it require any additions to your staff?"
- **As an attention-getter right before your conclusion:** "Now that we've examined the features of centralized word processing, what's the next step?"

8.5 HOW TO HANDLE QUESTION-AND-ANSWER SESSIONS

Many presentations close with a period during which you answer questions put forth by the audience. Some presenters revel in the opportunity to expand on a favorite topic, but others cringe at the thought of having to come up with a coherent response on the spot. View this session as an opportunity for genuine communication between you and the audience.

This "unscripted" part of the speech can be most illuminating for listeners, who now have the chance to (1) seek clarification of material you covered, (2) ask for additional information about the subject, or (3) disagree with points you raised. Here are a few guidelines for answering audience questions.

Let the Person Complete the Question

Speakers who are familiar with their material sometimes tend to anticipate questions and cut people off before they have finished asking the question. Besides being discourteous, this approach sometimes causes the speaker to make false assumptions about the question and, thus, to miss target on the answer. Let the audience member finish the question. In those rare instances when a person talks too long without zeroing in on a question, you can offer a mild corrective such as, "I'm not sure I understand where you're going with that question. . . ."

Be Sure You Understand the Question

Some questions are poorly phrased. If you're not sure what has been asked, request that the person rephrase the question. Another approach is to rephrase it yourself before you answer it. For example, you might respond by saying, "If I understand you correctly, you want to know if the insurance premiums are likely to go up more in the next five years than they have in the last five. Do I have that right?"

Repeat the Question Aloud

Listeners get frustrated when speakers begin answering questions that were not heard by the entire audience. If you have any doubt that everyone in the room heard the question, repeat it so everyone can hear it. Of course, you can simply ask if everyone heard the question, but be aware that some listeners will be reluctant to tell you even if they did not. It's best to repeat the question if there is any doubt.

Pause for a Moment Before You Begin Your Answer

This technique helps you gather your thoughts before you launch into an answer. The extra few moments are crucial for making sure you have an answer and creating a quick outline in your head. The pause also gives listeners the impression that you have carefully considered the question, as of course you have.

Admit It When You Don't Know an Answer

Most of us don't like to admit when we lack knowledge about a topic about which we have been asked to speak. But, invariably, you will encounter questions that you are not able to answer. In such cases, it is best to say right away that you do not have the information needed. Tell questioners that you will locate the answers later and get back to them. When you say you will follow up, of course, always do so.

Avoid Being Defensive

Sometimes you'll face questions or comments to which your natural inclination might be to respond defensively or to debate the point. If debating is the purpose of the event, then proceed. But most speeches are not debates, and a quarrelsome exchange only detracts from the impact of the speech. Do your best not to respond in kind to those who are trying to engage you in an argument. Instead, (1) acknowledge the differences of opinion openly, (2) agree to disagree on the point, or (3) offer to discuss the point after the speech.

Always End on Time

Follow the time limit of the Q&A period just as rigorously as you stay within the time limits for your speech. Simply because there are still questions being asked does not mean you should go beyond the time allotted. If there are still questions, you might want to suggest that the individuals contact you by e-mail or phone for further information.

8.6 *FINAL PREPARATIONS CHECKLIST*

Here is a checklist to help you make your final preparations for your speech. All presentations are different, so customize this list to meet your particular needs.

1. Presenter's notes and backup copies
2. Visual aids and backup copies. Laptop computer (with fresh batteries) or CD, Zip disk, or flash drive (sometimes called a *memory stick*). Have a set of paper copies as backup
3. Props
4. Professional clothing and shoes
5. Handouts printed out and collated. Take your handout master copy with you. Take extras.
6. Pointer or laser pointer (with fresh batteries)
7. Your business cards and contact information with the name and number of person who set up the presentation. This is the person with whom you decided ahead of time who will be responsible for printing and distributing handouts.
8. Address of, and directions to, the venue where you will be speaking. A map can help keep you from getting lost.
9. Transportation to the venue
10. Lozenges or cough drops
11. Spare clean shirt

TAKING IT TO THE NEXT LEVEL

While beginners should never skimp on practice time, once you have become a more accomplished speaker, you will discover that you need less practice before you speak. Presenters who give the same talk over and over sometimes find that they are getting bored and that their presentation sounds stale. Professional actors are able to perform the same show every night without getting stale by carefully preparing themselves mentally to get "up" for each performance. If you have to repeat a presentation more than two or three times, you might want to consult an acting coach or read up on acting techniques to learn how to do this.

12. Names and numbers of people you asked to help with demos and experiments
13. Pen, chalk, whiteboard markers, transparency pens
14. Stapler or paper clips
15. Watch or clock (with fresh batteries)
16. Microphone and amplification system (with fresh batteries)

ENCOURAGEMENT

Nonverbal cues are far more important than the words you use when it comes to retention. Put your energy and enthusiasm into your presentation and your audience will respond positively. Even if you aren't the greatest speaker, if you speak from the heart, your message will get through.

EXERCISES

1. Choose a partner and get a copy of a daily newspaper. Find the "Letters to the Editor" page and select one of the letters. Read the letter aloud to your partner, first in a dry, colorless, matter-of-fact voice, then with as much passion and emotion as you can muster. Have your partner do the same thing with a different letter. Discuss how it felt to speak with and without energy and enthusiasm.
2. Following are examples of the kinds of oral assignments typically given in college communication classes. Students are asked to prepare and deliver

a speech working within time limits. For each of these assignments, follow the guidelines in this chapter, especially those concerning speech notes and delivery.

- **Speech on Your Academic Major.** Prepare and give a presentation in which you discuss (a) your major field, (b) reasons for your interest in this major, and (c) specific career paths you may pursue related to the major. Assume your audience is a group of students with undecided majors, who may want to select your major. Use at least three graphics.
- **Speech Based on a Written Report.** Using a written assignment you have prepared for a recent course, develop a speech based on the report. Assume that your main goal is to present the audience with highlights of the written report, which you should assume listeners have not yet read. Use at least three graphics.
- **Speech Based on a Proposal.** Prepare a presentation based on a proposal you have written for a college course or for a job. Assume your listeners are in the position of accepting or rejecting your proposal but that they have not yet read the written document.
- **Speech Based on a Formal Report.** Prepare a group or panel presentation using the size groups indicated by your instructor. The speech may be related to a collaborative writing assignment or done as a separate project. In planning your panel, be sure that group members move smoothly from one speech to the next, creating a unified effect.
- **Team Persuasive Presentation.** Prepare a group presentation on a public policy issue. Have each member of the team research a different aspect and work together as a team to persuade your audience to change the way they feel, act, or think.

9 Evaluating Presentations

IMPORTANCE OF EVALUATIONS

When we get up in front of an audience, we want people to like us, to like the way we speak, and to value what we have to say. At the same time, we know that we are being judged on a wide range of factors—clothing, hair style, accent, facial expression, posture, and nonverbal messages are all being observed and evaluated when we speak. If we don't make a strong, positive first impression, audience members may decide they don't care for us and we may fail to achieve our purpose.

Without feedback, we will never know whether we have achieved our purpose or not. But, getting feedback entails the risk of being criticized. So, it is very difficult for some presenters to ask for feedback because receiving criticism can be painful. Nevertheless, getting an honest, informed critique of a presentation can be extremely valuable.

This chapter will describe why evaluations are important, how to collect feedback, and ways to use feedback and evaluations to keep improving as a speaker.

9.2 *ASKING FOR FEEDBACK RESPONSES*

Evaluation Tools

The three most common tools for evaluating a presentation are getting feedback from an expert observer, using paper audience evaluation forms, and making a video or audiotape recording.

Asking a "Friendly Observer" for Feedback

Ask a trusted person who knows about presentations to observe you and tell you how you did. While it may be hard to ask for feedback, getting constructive criticism from someone who is knowledgeable about public speaking can be a great help. The emphasis here is on *constructive* and *knowledgeable.*

Constructive criticism is designed to help you improve rather than focusing on what you did wrong. A constructive critic is someone who wants you to succeed and will help you understand what you are doing right and what you can do better. Similarly, a knowledgeable evaluator is someone who has some experience as a speaker and who understands the principles of good presentation skills described in this book.

Figure 9–1 is an example of a feedback form that focuses on how a talk was prepared and delivered, rather than on the content. It is used in a college class on technical presentations, but it could be adapted to any class on public speaking or oral communication skills. It lists some specific things you can ask a "friendly observer" to listen to and watch for when you give your presentation. This form might be used in a practice run before you do the actual presentation.

The following list of questions is much longer and more comprehensive, so you might want to have your friendly observer focus on just those that you find most challenging instead of trying to cover them all:

- In the introduction, did you establish your purpose and explain your credentials? Did you start with some kind of audience "hook" to get audience members interested in your topic?
- Did you forecast what your presentations would be about and explain the format you would follow?
- In the body of your talk, was the flow of ideas logical and transitions between parts smooth?
- In the conclusion, was there a summary of your talk? Did you wrap up with a solid conclusion and a call for action, if appropriate?
- Did you finish your talk within the time limit? Did you use your time effectively and efficiently?
- During the question-and-answer portion, did you repeat each question before answering it?
- Was your voice expressive?
- Was your voice loud enough to be heard by all?
- Was your pace appropriate and not rushed?
- Were you articulate? Did you enunciate clearly? Were all words pronounced correctly and clearly?
- Did you communicate with energy? Did you show enthusiasm for your topic?
- Did you speak with your head up, facing the audience and making eye contact?
- Was your appearance professional-looking—neat, well-groomed, and appropriate for the occasion?
- Did you come out from behind the lectern—move and gesture?

Speaker: _____

Topic: _____

Category	Very Good	Good	Acceptable	Poor	Very Poor	Comments
Organization (shows good choice of topic, logical development, clear opening and closing, good time management and pacing, comfortable interacting with audience's questions and comments)						
Voice (clear, confident, loud, with steady pace and showing presenter's enthusiasm for and knowledge of the subject)						
Physical Presence (shows speaker's ability to interact with and to engage audience; good use of non-verbal communication)						
Language (level and terminology appropriate to the audience; explanations clear, concise, and logical)						
Illustrations (creative, legible, relevant, and support theme; a good mix of text and graphics, simple design)						

FIGURE 9–1
Sample feedback form

- Were you able to come across as a warm, friendly, credible presenter? Did you engage your audience and establish some level of rapport with them?
- Was your choice of language appropriate for your audience? Did you speak at the right level of formality or informality? Did you define technical terms and avoid jargon and slang?

- Were you able to avoid filler words (*uh, OK, basically, like*)?
- Were your sentences short, simple, and declarative?
- Did your visuals work for you to enhance your oral presentation and make information more accessible?
- Were all sources cited?
- Were you able to interact with your visuals?
- Was your use of humor appropriate?
- Did you accomplish your goals (to inform, persuade, train, or entertain)?
- Were you prepared? Was it apparent that you had practiced and that you did your best?

Paper Evaluation Forms

Figure 9–2 is an example of an Audience Evaluation Form to help you understand how the audience responded to a talk. It would be very easy to adapt this particular form to any kind of speech. Note that it is easy to fill out, that it encourages respondents to say something positive about the presentation before they criticize "things that could have been better," and that it provides a way for audience members who have questions for the speaker to get a personal reply.

Paper evaluation forms can be a useful way of getting feedback from the audience. Some presenters find reading these forms difficult because they don't handle criticism well. You can make it easier for yourself to read and use evaluation forms if you frame questions carefully and follow these guidelines:

Keep Forms Simple

The easier you make it for audience members to fill out an evaluation form, the more likely you are to get it back. The two main factors that keep people from filling out evaluation forms are time and complexity. Therefore, you need to design a simple form that can be filled out quickly and still yield valuable information. This can be a real challenge, for if it is too simple, the information you get back will not be worth much. And, if it is too complex, people will resent having to spend a lot of time figuring it out and writing in answers.

Ask Questions That Can Be Easily Answered

Stick to short-answer questions that call for *yes/no* answers or ask people to rank responses on a simple scale from Very Much to Very Little. Wherever possible, give them boxes they can check off. Include an *Other . . .* category for those who want to tell you more.

Leave Space for More Detailed Answers

If you want to get more details about the speech, leave space below the short answer questions for audience members to write comments.

Speaker: _____

Topic: _____

Please complete this form and turn it in to the speaker at the end of the presentation. Thank you in advance for your comments.

	Very Good	Good	Poor	Very Poor	No Opinion
1. The talk covered the material well					
2. The speaker was well-prepared					
3. The speaker was knowledgeable about the topic					
4. The visual aids helped me understand the topic					
5. The handouts helped me understand the topic					
6. The presentation made sense and was easy to follow					

7. The best thing(s) about this presentation was (were): _____
_____
_____

8. Thing(s) that could have been better about this presentation: _____
_____
_____

If you have a question for the presenter and would like a personal answer, please complete this section:

Name: _____

Your question: _____

Phone: _____

Email: _____

FIGURE 9–2
Sample audience evaluation form

ANONYMOUS EVALUATIONS

Many people believe that, in order to get honest opinions on an evaluation form, the evaluator must remain anonymous. While there may be some truth to this, anonymity makes a dialogue between speaker and audience member impossible. It eliminates any possibility of follow-up, clarification, or sharing of resources. If you have a choice, consider providing space on the form for the evaluator's name.

Encourage the Audience to Respond

You can't really blame busy people from wanting to leave as soon as your presentation is over. This can make it a challenge to get them to fill out and turn in an evaluation form. Some presenters do this by offering a reward (a piece of candy or a handout packet) for filling in the form and turning it in. Others try to persuade the audience that their opinions matter and that they value the feedback they get.

One suggestion for encouraging audience members to return evaluation forms is to print them on brightly colored paper. Sometime near the end of your talk, distribute these forms and explain what you want done with them. It will be much harder for audience members to walk out without turning in their forms if the forms are so eye-catching.

Allow Time to Fill Out Evaluations

You are more likely to get evaluation forms filled in if you allow audience members time to work on them. This should be part of your allotted presentation time, not something you tack on at the end. Don't rush them or you won't get good feedback.

Offer to Provide More Information

You may use evaluation forms as a way of continuing your presentation with audience members. Provide a space on the form where they can ask a question and fill in a name, phone number, and e-mail address where you can reach them with an answer.

Include evaluation forms with the handout packet you give audience members. Identify what they are near the end of your talk and explain what you want them to do with them.

Videotape Is a Great Way to See and Hear How You Did

You can learn a lot about your strengths and weaknesses from watching yourself present on videotape. Here are some tips for videotaping yourself doing a presentation:

- Find out ahead of time if your talk will be recorded. If so, ask for a copy of the recording. If not, arrange to have someone operate a video camera during your presentation.
- Talk to your camera operator ahead of time to that he or she understands that you are interested in seeing how you look and sound in front of an audience, so the camera should focus on you.
- Mount the camera on a tripod so it stays steady and raise it as high as you can so you can be seen over the heads of the audience. Set it up behind the audience so that it doesn't interfere with their sightlines.
- Keep the lighting in the room bright enough so that you can be seen clearly in the camera viewfinder.
- Video cameras have built-in microphones, but if you use a microphone connected to the camera you will get a truer sense of what your voice sounds like. When you play back the recording, listen as well as watch yourself; hearing how your voice sounds is also important.
- Treat the camera as if it were a member of the audience; don't stare at it, but don't ignore it entirely either.

Watch the videotape of your presentation carefully. Try watching it once with the sound off, so you can focus on your nonverbal communication, and once with the picture off, so you can listen to your voice. You might want to have your "friendly observer" watch the tape with you to point out what you did well and what you still need to work on. Once you have tried using audio or videotape to record your presentations, you will know how valuable they can be.

Using the Feedback You Get

The most important reason for asking for feedback is a sincere desire to learn from your successes and failures. You can ask a "friendly observer" to give you feedback or you can ask audience members for their reactions to your talk. Once they have given you their opinions, you have to decide whether and how to act on them.

Audience responses to speakers tend to fall into the normal distribution pattern that scientists and mathematicians call a *bell curve*. At one end are a small group of folks who respond with such strong positive feelings about you that, in their eyes, you can do no wrong. Don't worry about them; they are already on your side. At the other end are those few who dislike you the minute you start talking (or before!)—they hate your clothing, your haircut, your accent, the organization you represent, your topic, and so on. Perhaps they don't really want to be there at all. Don't worry about them; there is nothing you can do or say that will change their minds.

The vast majority of audience members, however, are people who have no opinion about you or your topic until you step up to the lectern and start speaking. As you read their comments, remember that you prepared as best you could and that you had valid reasons for the decisions you made about your speech of which others are not aware. So, take comments more or less seriously, but consider the source before you get upset or disheartened.

Some criticisms should be given less weight because they deal with things over which you have no control. Complaints about the seats being too hard, the room being too hot, or the air conditioning being too loud have nothing to do with your speech and should be turned over to the building's manager.

Other comments about the content and structure of your talk should be given greater weight, because they may indicate a problem with your audience analysis or your organization and delivery. But unless you have completely botched your speech preparation, use what you read on evaluations to fine-tune your skills and not as a reason to revamp your entire wardrobe, personality, voice, and lifestyle.

Applying Lessons Learned

Here are some tips on how to keep developing your presentation skills.

- Make a "For next time" list. Keep track of any possible changes you might make the next time you give a talk. Make it a habit to check your "For next time" list every time you are scheduled to give a talk.
- Don't make changes just for the sake of making changes. If your presentation went over well and the majority of the audience turned in positive evaluations, think twice before deciding to do things differently. If you do decide to make a change, make it the smallest change possible— don't revise an entire presentation just because one definition wasn't completely clear.
- Delete instead of add. New presenters sometimes make the mistake of trying to pack too much into the time allotted. If you decide to make any changes the next time, consider cutting things out rather than adding more. A pruned tree will grow better.
- Keep track of what works; compare the evaluation responses to the purpose you defined when planning your speech. If you accomplished your purpose, don't touch anything—"If it ain't broke, don't fix it!"
- Make note of the questions you were asked by the audience so you can include the answers in the body of the presentation next time you give it.

9.3   *OBSERVING TO LEARN*

You can learn a lot about what works and what doesn't in public speaking by observing other people's presentations. Try to take opportunities to observe presenters in your field whenever you can. Pay attention to what they do and how they do it. Focus on those things that affect their credibility, their rapport with the audience, their use of examples and visual aids, and their style of delivery. Learn from other presenters' successes and failures

ENCOURAGEMENT

Not everyone will adore you, but most of the people in the audience will want your presentation to be a success. They want you to be interesting and credible. Use constructive feedback to help you improve the way you come across.

EXERCISES

1. There are many different ways to evaluate the success or failure of a presentation. Some prefer to distribute evaluation forms to audience members. Others like to watch themselves on videotape and critique their work based on that. Write a short essay discussing some of your ideas about how presenters can evaluate their work more effectively.

2. After you do a team presentation, set up a team meeting to review your experience and to evaluate your work. Discuss how the work was divided

---

TAKING IT TO THE NEXT LEVEL

You will continue to improve only if you keep working on your presentation skills. Take advantage of any openings you get to do presentations. Join an organization, like Toastmasters, which offers opportunities for public speaking.

In many workplaces the job of taking schoolchildren and other visitors on a tour of the company is considered undesirable and delegated to those with less seniority. This foolish attitude offers anyone who is interested in polishing their presentation skills a wonderful opportunity: You'll get to learn a lot about how your company works, you'll meet more of the people who work there, and you'll become an expert at talking to nontechnical audiences about your work.

and whether you feel that the division of labor was fair. Here are some questions to consider:

- Did you accomplish your goals? Were you informative, persuasive, instructive, or entertaining?
- Comment on your delivery skills, use of visual aids, voices, and interactions with the audience.
- What was it like preparing and presenting as part of a team?
- Comment on each person's participation. Did everyone do his or her fair share?
- What would you do differently if you were to do it over again?

3. One of the most difficult situations for team members is when there are noticeable differences in the quantity and quality of work contributed by members of the team. This can result in hard feelings, especially for those who had to do more than what they feel is their fair share. What can teams do to overcome this problem? Graphics, like those shown in Figure 9–3, can be used to prompt a discussion about how team members felt about the division of labor. The pie chart on the left shows that the members of Team Wonderful felt that everyone contributed equally. The chart on the right shows that some members of the Slackerz Team did more than others. Write a short essay discussing fair ways of evaluating the quality and quantity of participation by team members.

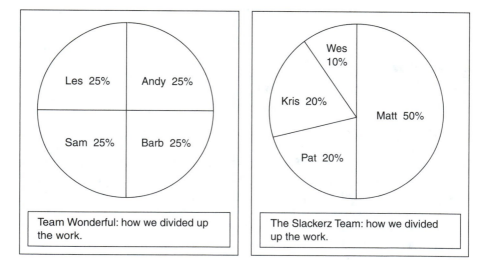

FIGURE 9–3
Graphic showing team members' participation

10 Adapting to Different Situations

Other texts on public speaking and technical presentations tend to focus on traditional speaking situations where you are asked to prepare and deliver a speech. In the world of business and industry, however, there are many more situations to which you will need to adapt your speaking skills. Here are some common situations in which your skill at delivering a speech can lead to success for you or your organization, or both:

- **Getting hired.** A hiring committee may ask you to give a short talk summarizing your education, experience, and career goals.
- **Getting customers.** A potential client may ask you to speak on the highlights of a large proposal you recently submitted.
- **Keeping customers.** A long-term client may ask you to give a speech to the client's employees on technological advances in your products or services.
- **Contributing to your profession.** You may make a speech (or "deliver a paper") at a meeting of your professional association.
- **Contributing to your community.** You may be asked to give a speech on your profession to a community group.

While it would be impossible to cover every single situation in which a person might have to speak, this chapter offers advice on how to handle those that are most important to your career. These include workplace situations—doing a persuasive presentation of a proposal or a bid, conducting an on-the-job training session, and running a public meeting.

10.1 SPEAKING ON THE JOB

Your oral communication skills can lead to success for you or your organization, or both. In order to get your career started, though, you will have to go through the process of interviewing for a job, which is also a form of presentation.

Getting Hired

Job Interview

Although the Internet has had a big impact on the way people search for and apply for jobs, one aspect of the process remains the same—the face-to-face job interview. Employers want to meet applicants in person, because they consider this a valid way of measuring character, personality, and interpersonal communication skills.

Some applicants who "look good on paper" do not come across well in interviews. To persuade an employer that you have the right combination of skills, you should think of the interview as a form of persuasive presentation. When you are asked by an interviewer to "Tell me about yourself," don't merely repeat information that can easily be found on your résumé—use it as an opportunity to do a mini-presentation that will showcase your communication skills and go beyond the bare facts found on your résumé.

Most job-hunting today involves a process in which the applicant submits a letter describing his or her background and skills attached to a résumé summarizing the applicant's experience. The next step is usually a live interview in which a representative of the employer meets with the applicant. The high cost of travel and the introduction of new technology have led some companies to replace or add to the traditional face-to-face interview with a telephone call or a video-teleconference. For information on how to handle this situation, see "Tips for Videoconferencing" later in this chapter.

In a typical job interview, the interviewer or hiring committee may ask you to give a short talk summarizing your education, experience, and career goals. You may wonder why they do this, because the employer's representative will have read your résumé and cover letter before the interview and already knows about your background, he or she wants to talk to you in person.

Here's why. Experienced job seekers know that the real test comes after an interviewer has gone over their official qualifications—when he or she looks up and says, "So, tell me about yourself." What they are asking for is a mini-presentation (for a tiny audience) to see how well you can communicate.

Because you can count on being asked this question, you can prepare your answer ahead of time. Here are some tips on how to handle the "So, tell me about yourself" question in a job interview.

- Do not repeat information that is on your résumé or in your cover letter.
- Use this as an opportunity to expand on your résumé: Pick one or two of your strengths in advance and talk at length about how things you have done and learned can be of use to the employer. Treat this just as you would any presentation—focus on your delivery skills, organization, and clarity.
- Go into detail about your skills and experiences and show direct connections to the position you are applying for.

- Talk the talk: Use the vocabulary of someone in this field. Practice using the terminology until you sound like you know what you are talking about.
- Do your homework: Read up on the company so you can show how useful you would be as an employee. This will also help you to ask insightful questions about the organization and what your role will be in it.

Showcase Your Communications Skills

How well you handle the interview will determine whether you get to go any further in the hiring process. If you think of the job interview as a test of your communication skills, you will understand why you can't stop there.

Thank-you Note

Consider following up your interview with a written thank-you note. Put something in the note that will help the recipient remember who you are and add to (or reinforce) what you said in the interview.

Telephone Etiquette

Another communication skill that sometimes gets overlooked is your ability to use the telephone. Here are some tips for phone use in the workplace:

- Whether you are making or receiving a call, always start with "Hello" and identify yourself to the person on the other end of the line.
- Always use "Please" and "Thank you" on the telephone. You may be grateful or happy, but the other person can't see the smile on your face, so make sure they can hear it in your words.
- When engaged in a business call, give it your full attention. Turn off the TV, radio, computer, or CD player.
- If you leave a message, make sure you speak slowly and clearly. Spell out your name and say the number where you can be reached twice—slowly and clearly so the other person has time to write it down.
- When you record your outgoing message on your answering machine or voicemail system, speak slowly and clearly. Keep it clean and avoid slang or jargon.

Leading a Plant Tour

Whether you are showing off to a group of investors, a class of seventh graders, or the corporate brass, you need to know how to present ideas clearly and concisely. This is an important presentation; unfortunately, in too many companies this job is considered undesirable so it is foisted off on the least experienced person around—often someone who has only recently joined the company and doesn't know much about things.

DIRTY WORDS DON'T CUT IT

A few years ago, one of my students spent a semester as an intern at a large company. One of the jobs he was given was to contact a number of job applicants to set up job interviews with the intern's supervisor.

During the course of one such call, the intern started laughing loudly enough to attract the supervisor's attention. When the supervisor asked why he was laughing, the intern handed him the phone and dialed the number of one of the job applicants. A machine answered the phone and played a recorded comedy routine from a TV cartoon that featured a lot of dirty words.

The supervisor listened to the message and hung up the phone. The intern was told to skip to the next name on the list, "That's not the sort of person we are looking for," the supervisor told him.

A plant tour consists of showing visitors your company's facilities, introducing members of the staff, and explaining the various production processes. A good way to understand what you are trying to do is to consider this an oral "process description" that will be illustrated by the real thing.

Here are some tips for handling a tour of your workplace:

- Do a thorough audience analysis. Find out in advance who your visitors are, why they are visiting, and what is important to them. For example, for a business audience, stress safety, efficiency, maintenance, and quality controls. For schoolchildren, talk about the importance of education and training to being successful in your industry.
- Prepare information handouts in advance, so the audience can see data while listening to you. Your handout might include a flowchart or map that visitors can follow during the tour.
- Explain to your visitors what they will be seeing before they enter noisy areas. Try to talk where it is quiet.
- Be careful to define technical terms and avoid jargon.
- Know the names and titles of the people you will be introducing on the tour.
- Get the names of the machines and equipment you will be showing your guests; visitors often ask questions about where machines come from, how much they cost, how fast they can go, how they compare to other makes and models, etc.
- Point out safety and quality control measures wherever appropriate.

Presenting to Your Peers

Staff Meetings

Most organizations today have some kind of regular staff get-together—usually weekly—to share news and important information. These are sometimes used to distribute orders and information from management, but it is also common for them to involve an exchange of ideas among staff members. If you are sent to a training session, attend a conference, or participate in a meeting of a committee you may be asked to summarize your experience for your co-workers. Therefore, it's important to learn how to present information to one's peers.

Presenting

Here are some tips for presenting at staff meetings:

- Be concise, make it clear why the information is important, and don't waste time.
- Excuse people who are not affected by your presentation topic.
- Provide a handout with a bare-bones outline so the staff can follow along and learn the terminology.
- Allow time to discuss any follow-up to your report. Tell people if you plan to do something and conclude with call for action on their part if it is appropriate.

Facilitating

Facilitating meetings is beyond the scope of this book, but you should know that running meetings is a skill that can be learned. Your library and bookstore have many books on how to chair a meeting effectively. Here are a few thoughts about the importance of communication in facilitating meetings:

- Set up and publish an agenda ahead of time. Make sure everyone knows the reason for the meeting and any ground rules that apply.
- Time is valuable—don't try to do too much at one meeting.
- Don't end the meeting without deciding on the next steps.
- If you want to gain a reputation for being a great facilitator, start on time, end early, and speak no more than 10 percent of the time.

Addressing a Professional Association Meeting

At some point in your career, you may be asked to give a speech or "present a paper" at a meeting of your professional association. Your invitation to speak depends upon your being able to contribute to the knowledge of others in your field. Academics, for example, are expected to discuss their research. Other professionals are invited to comment on trends and to offer insights into contemporary issues.

Presenting an Academic Paper

If you are affiliated with a college or university, you may be asked to submit an abstract—a summary of an academic paper—in advance. The organizers of the conference will read the abstract and then decide whether they want to invite you to present your paper at the conference. It has long been the tradition in academia for those presenting papers to read them verbatim to the audience. The readings are followed by question-and-answer sessions.

Some academic disciplines never use visual aids; others use them all the time. If you are asked to deliver a paper, find out if you will be expected to read your paper verbatim and if you are supposed to use visuals. In some cases, in addition to presenting your paper, the organizers may want you to participate in a panel discussion or a poster show. Both of these are very popular at academic conferences.

Preparation for presenting at an academic conference should consist of writing a clear, well-organized research paper and knowing enough about the topic to handle questions at the conclusion. It is common to find half a dozen presentations going on at once at these conferences, which means that attendees have to pick and choose which ones they will be able to observe. Some conferees attempt to cover as many as they can by popping in to catch five minutes of a presentation, grabbing a copy of the handouts, and slipping out to do the same thing at the presentation next door. You can make up a short handout for these casual observers who don't stay around for the whole presentation and leave a stack of them near the door.

Addressing a Professional Association meeting

Most professionals are members of professional associations that promote their interests, present their views to the public, and lobby for favorable legislation. Doctors, lawyers, librarians, accountants, social workers, nurses, and members of many other professions belong to these associations. Local chapters of professional associations often have monthly meetings to which you may be invited to speak. Regional and national organizations usually meet once a year. Some professionals are required to attend training workshops to earn Continuing Education Credits (CEUs).

While different conferences may follow different formats, the following situations are common to many.

- **Keynote address.** The first or last speaker is sometimes referred to as the keynote speaker. The keynoter is expected to give a "podium speech" in which he or she addresses a broad, important issue related to the profession. Sometimes the rest of the conference will be devoted to workshops discussing the issues raised by the keynoter. For a podium speech, prepare and practice from a written script with the exact words you want to use. Keynote speakers do not usually use visual aids or interactivity. If time permits, they may answer some audience questions at the end of the speech, but, because audiences are frequently very large for keynote addresses, this is difficult to do. If you are a keynoter and you would like to interact with members of the audience, ask the conference organizers if they will arrange for a reception or a small discussion group at the conference.

- **Breakout sessions.** These often begin right after the keynote speaker has discussed the main theme(s) of the conference. There will be several individual speakers or panel discussions taking place simultaneously and conferees will have to decide which they want to attend.
- **Panel discussions.** A group of people with some expertise on a topic will come together as part of a panel to discuss that topic. Panels often have a moderator who can act as a facilitator to keep the participants on task. Most of the time, each panel member will be given a set time to discuss what he or she knows about the topic. This is often followed by a discussion by the panel members, in which they respond to one another. Then the audience is invited to participate by asking questions of the panel. Here are some tips if you are asked to be on a panel:
  1. Get the names and contact information for the other panel members and get in touch with them beforehand to work out what each of you will each be discussing. You will save yourselves a lot of time, trouble, and potential embarrassment by doing this ahead of time.
  2. Send the other panel members a short summary of what you plan to talk about.
  3. Prepare handouts for audience members describing your ideas. These are helpful for those who have to leave early.
  4. If you are concerned about keeping the content of the panel discussion focused, make up a short list of questions and give it to the moderator. Explain that these are issues he or she can raise if the panel discussion starts to wander or if audience attention begins to lag.
- **Workshops.** Workshop attendees are there to learn skills that they can apply to their profession. If you are asked to run a workshop, plan to do an instructive presentation, which may involve interactive exercises and activities. Make sure you are clear about the learning objectives for the workshop. For more on training people, see the discussion in this chapter.
- **Poster shows.** If you have been invited to speak about your research, you may also be asked to create an exhibit for a poster show. These posters are static displays, usually two-dimensional, that illustrate the parts of a research project. They can be produced in so many ways that it is difficult to generalize. Some presenters make paper copies of Power-Point visuals, tape them to a display board, and call it a poster. Others have posters designed by graphic artists, printed out full size, and laminated. A lot depends on how much time and money you have to put into your poster.

From a presentation point of view, the important thing about poster shows is that time may be set aside during the conference for participants to stand next to their visual displays and to explain their work to attendees. Here are a few tips for talking about your poster show:

1. Limit your talk to main points; don't go into details.
2. Be prepared to repeat the same talk several times to an audience that is standing near you reading your poster. Keep your presentation very short—about two to three minutes.
3. Let your poster do most of the work. Use it as a visual aid and point to the images as you talk about them.
4. Try to get a sense of the other posters in the show, so you can explain how your work relates to others'.
5. Bring a stack of business cards with you so people who have questions can get in touch with you later.

10.2   PERSUASIVE PRESENTATIONS

Progress and change in the worlds of business, industry, and academia often depend upon people's ability to recognize opportunities when they arise and to muster the support necessary for meeting them. While most changes are initiated with a document—a proposal or business plan—a key part of the process is an oral presentation. In these situations, a presenter must know how to change the attitude, behavior, or thinking of the audience. Following are some examples.

Proposals and Business Plans

Many projects are initiated by a written proposal that explains:

- What can be changed and why it should be changed
- How the proposer thinks it should be changed
- How long it will take and how much it will cost
- Why the proposer should be given time, money, and other resources to implement the proposal

A complete explanation of how to organize and write a proposal can be found in any text on technical writing. A business plan is a proposal for starting a new business enterprise or for redirecting an existing business. Here are two examples of persuasive presentations of proposals:

- **Getting customers.** A potential client may ask you to speak on the highlights of a large proposal you recently submitted. (In engineering, these are sometimes referred to as "interviews.")
- **Keeping customers.** A long-term client may ask you to give a speech to the client's employees on technological advances in your products or services.

Guidelines for Proposal Presentations

Here are some guidelines to follow if you find yourself doing a persuasive presentation of a proposal.

- Stick to those issues of interest to your audience; disregard your own feelings and priorities.
- Work within your professional ethics. Limit yourself to rational persuasion and avoid anything that might be construed as manipulative, biased, or misleading.

Explain why, in your professional opinion, the solution you recommend is the best one for your audience.

- Don't disparage your competitors.
- Your audience analysis must include determining the appropriate level of sophistication. Talking over their heads is as bad as talking down to them.
- Hand out a glossary if there are a lot of technical terms.
- Do not assume that audience has read your document. Explain everything clearly and define terms.
- Some presenters think that if they can show enough data, the audience will be persuaded. The result is that instead of having two clear, easy-to-understand visuals, the audience has to sit through ten. This data-driven approach is not terribly persuasive—instead of getting them fired up over your proposal, it may, in fact, bore your audience to tears. Instead of weighing them down with data, tell stories and give examples. Use pictures to help them understand your approach to the situation. They will remember your stories much longer than they will the numbers and statistics.
- Given forty-five minutes, many presenters will talk the entire time in the mistaken belief that the more they talk, the more the audience will be persuaded. You might do better to spend more of your allotted time listening to your audience. Listening can be especially effective in a situation in which audience members want to discuss the information as it is presented. People expect you to give them information, but then they like to discuss it. If you do it right, they will persuade themselves.

HOW TO PRESENT A PROPOSAL

There is some disagreement among experts about the best way to present a proposal. Some argue that your oral presentation should come directly from the written document. Others say that the two need to be different. Here is a summary of the arguments in favor of presenting a proposal exactly as written:

a. Audience members may not have read the document. Readers are too busy to read an entire proposal and they are frequently interrupted. They also only read the parts that pertain to them, so they don't really know what is in the rest of the written proposal. And, some audience members may be poor readers or illiterate.

b. If the written proposal is organized and planned, why change it? Assuming that the document has been researched, the material should be well organized, and the explanations have been polished, why do anything different?

c. The written proposal is already persuasive, so why take a chance of confusing people by making the oral presentation something different? The words and phrases in the document were specifically chosen to sway readers, so why risk changing their meanings or hurting the impact of proposal?

Here are the arguments in favor of making the oral presentation different from the written proposal:

- Prime the pump: Put questions you want your audience to think about on your visuals. Frame the questions so that your audience will come to the conclusion that your proposal is the best way for them to reach their goals.

Team Presentations

Proposal presentations are sometimes done by a team of presenters. This is especially true when the proposal has been put together by a team of people from one or more companies. Coordinating a team presentation can be difficult, more so when the presenters are working in different fields or for different companies. Here are some tips for doing presentations as part of a team:

- You should all agree ahead of time who will say what.
- This may seem obvious, but all of you should know what is in the proposal. Not only will you avoid embarrassing yourselves, but if one mem-

a. Readers of the written proposal can decide for themselves how much time they will spend reading and they can go back and reread parts of a document. Presentations are usually time limited, so there is no opportunity to review or go back. Readers can decide to read the entire proposal or just the parts that concern them, while the choice of what to focus on in an oral presentation is up to the presenter. You need to have good reasons to cover material that is in the document if you have limited time and need to sell your ideas.

b. The oral presentation is really a sales pitch for the idea of the proposal, not for the document. A technical document, such as a written proposal, tries to cover everything, so readers can find what they want, while a presentation can focus on what is important.

c. The organization of an oral "sales pitch" differs from that of a written proposal. When trying to persuade, you start by identifying main selling points of your product: what are the strongest benefits to your audience? Then you rank order these items and focus your presentation on the most important advantages of your plan. You don't bring up weak points or disadvantages, but you must prepare to meet objections as they come up.

Which strategy you should use really depends on your audience. If they have all read your proposal and know what is in it, go with a sales pitch. If they have only read part or none of it, you need to explain what is in the document and then go into your sales pitch.

ber of the team can't make it to the presentation, another team member can cover for him or her.

- Practice together if at all possible.
- Make one person responsible for visual aids, so they are consistent and professional.
- During the presentation, team members who are not speaking should sit quietly watching the presenter. A team member fidgeting or playing with the computer will distract the audience and hurt the team's credibility.
- Don't interrupt your teammates and don't point out any mistakes they make to the audience unless the error is material to the presentation.
- Find out ahead of time how long they expect you to talk and how much time will be taken up with questions.
- Make sure everyone looks and talks like a professional. Be confident and work to make your confidence obvious to your audience.
- Don't make promises unless you have the authority and intend to keep them.

10.3   *TRAINING*

Every good manager knows that having a well-trained staff can make all the difference in the success or failure of a business. The ability to train others could be an important skill in terms of your career. Technical experts are often called upon to explain and demonstrate new technology. Managers are responsible for explaining policies and procedures to their staffs. (To review instructive speeches, see Chapter 3.)

Guidelines for Running a Training Session

Here are some guidelines to follow if you are asked to set up and run a training session:

Set Learning Objectives

Decide in advance what you want your audience to be able to do after training is done. Explain the purpose for the training to the audience at the beginning of your presentation.

Practice

Practice on someone to make sure you know what you want to say beforehand. You may discover that something you thought was obvious or foolproof isn't and that you need to polish your presentation.

Don't Try to Do Too Much in One Session

Add more sessions if necessary and be prepared to repeat ideas in different ways if trainees don't understand the first time. Learning can be hard work—give trainees frequent breaks.

Avoid Being Patronizing or Talking Down to Trainees

Explain processes step-by-step in logical order. Define terms and processes and use the same terminology throughout training. Praise trainees when they get it right. Be encouraging and positive.

Try Not to Lecture

Get trainees involved from the very start. Ask trainees questions to measure learning. Stop at the first signs of confusion and explain again.

Be Patient

Stop frequently to ask if there are questions. Provide troubleshooting guides and resources.

Follow Up

Do an evaluation of the training at the conclusion of your presentation and try to do a follow-up a few weeks later to find out if your training had an impact.

10.4 PUBLIC MEETINGS AND HEARINGS

Technical experts are often called upon to explain to the public what they want to do (or have already done). Accountability is extremely important. This section reviews some things to remember when you speak as a representative of your company or profession.

Public Meetings

You may be asked to speak about a project or your profession to a community group. Depending on the circumstances, audience members may be sympathetic or hostile, so you need to be prepared for a wide range of responses. Here are some guidelines for speaking to a public meeting:

- Maintain a high standard of behavior and ethics. Express sympathy with critics without admitting responsibility, unless you intend to do so.
- Make sure you have the authority to speak for your organization before you address a public meeting or speak at a hearing. If you apologize for something your organization did, do so sincerely, quickly, and directly.
- Don't make promises unless you have the authority to do so and you intend to keep them.

Difficult Audiences

On occasion, you (or your organization) may be the one to organize a public meeting. The audience may be friendly, neutral, or hostile. Dealing with problem audiences can be difficult, so follow the guidelines for speaking to a public meeting and consider the following pointers:

- Decide first if a public meeting is the best way to deal with the issue. If only a small number of people are affected, you might do better with a small group meeting or even talking to them one-on-one.
- Make it personal if possible—greet audience members at the door when they arrive.
- Don't shy away from tough topics. If you have a sense of why your presentation might be controversial, think up answers to hard questions and practice them.

- Set an agenda and post it prominently. This will make it easier to keep people on track. You might even want to give out copies of the agenda as a handout to help people remember what was discussed.
- If things get heated up, give the audience a ten-minute break—enough time to cool down.
- If you know where hostile audience members are seated, try to alternate their questions and comments with those of people who are neutral or who favor your position. This will keep the meeting from turning into an endless series of complaints and criticism.

Testifying at Public Hearings

Legislators, from city council members all the way up to U.S. senators, collect information about public policy issues by holding hearings to which experts are invited to testify. The expert usually starts by reading a prepared statement to the committee and then answers questions about his or her position. Here are some tips for testifying at a public hearing:

- If you are called to testify before a legislative committee, you should prepare your statement in advance and run it past a lawyer. You might also prepare a statement for the media describing your testimony and containing contact information in case they want to interview you.
- Be extremely careful when you testify under oath. Your sworn testimony can be used against you and your company. If you do not tell the truth, you may be found guilty of perjury.
- Stick to one or two main points and be brief. Submit documentation to support your arguments separately.
- If your goal is to persuade, remember that politicians are moved by arguments that relate to their constituencies and their causes.

10.5   USING NEW TECHNOLOGY TO COMMUNICATE

Teleconferencing Background

One of the most common uses of new technology for communicating is teleconferencing—sending audio and video signals from one site to another. The sites can be across town or on the other side of the globe. One or more sites can receive the same teleconference.

Conference calling—linking two or more people on a single telephone conversation—has been around for a long time. The addition of video is somewhat newer, but it has spread rapidly since the terrorist attacks of 9/11, when many companies cut back on business travel. The cost of systems has come down, too, so even a small business can now afford to buy a basic videoconferencing system. As a result, the number of business meetings held via video- and audio-teleconferencing since 9/11 has skyrocketed. This technology is also widely used in distance education, where the teacher and students are miles apart. (Another twist is the ability to use the Internet for video conferencing. While sounds and images might not be as clear, the costs are so low that this is an extremely attractive option.)

Tips for Videoconferencing

People who aren't given some training before trying to talk to a camera and microphone don't realize how bad the electronic media can make them look and sound. Whether you are doing a videoconference or appearing on television, bad hair, annoying habits, thick accents, fidgeting, and mumbling can all make you look incompetent. Some of these things you have no control over, but there are many steps you can take to look and sound good in front of a video camera. Here are some guidelines for doing a video teleconference.

Arrive at Your Videoconferencing Site Early

- This will give you time to get ready for the session.
- Inspect the equipment to make sure it is in order.
- Adjust seat height and arrange other furniture the way you want it to be set up.
- Check out the lighting. Fluorescent lights can make presenters look sickly. Strong spotlights directly overhead will cast unflattering shadows on presenters that will make it look like they have dark circles under the eyes, a moustache, or worse.
- Locate the clock, so you know when it's time to give the audience a break and when your time is up.
- Use the rest room beforehand, if necessary.

Dress Appropriately

- Wear simple, solid colors. Clean, light-colored clothes look best. Avoid busy patterns such as those on Hawaiian shirts, hound's-tooth, or checks. On video they tend to create a distracting, stroboscopic (swimming) effect.
- Don't wear bright red and stay away from wearing all-dark or all-white. Some automatic cameras will be fooled by these colors and adjust so that you either look pale and washed out or like you are sitting in the dark. If you wear a suit, combine a dark jacket with a light shirt to keep from fooling the camera.
- Never, ever wear a hat on video (except for religious or medical reasons).
- Keep jewelry to a minimum. Save glitter earrings and gold chains for the discothèque. They can reflect light into the camera lens and distract your audience.

HOW VIDEO TELECONFERENCING WORKS

Technology has come a long way since two-way video was first demonstrated at the New York World's Fair in 1964. Once accessible to only the wealthiest corporations, it now costs less than $2,000 to set up a studio. Some companies avoid the costs of setting up their own systems by renting studio time from teleconference service providers. The cost of teleconferencing is determined by the equipment you buy, the transmission system you choose, and the cost of labor for setting it up and operating it.

Equipment

A typical video teleconference setup consists of a video camera that is pointed at the presenters, a document camera (sometimes called an *overhead camera*) camera that can be used to show printed pages, and a large video monitor to display pictures coming from the other connection. A microphone may be connected to the camera or set up on the presenter's table.

Fancier set ups may also include a separate Internet connection for showing websites as visuals, a DVD or VHS player for showing videos to your audience, and additional cameras pointed at the audience.

Connections

All of your equipment needs to be attached to a system that can send and receive signals from the other end. Most videoconferencing today uses one of two types of connections (called *bridges*): Integrated Services Digital Network (ISDN) or Transmission Control Protocol/Internet Protocol (TCP/IP)—the same networking system used by the Internet. This means you can choose

---

- Eyeglasses are okay.
- Get a good haircut. Men should shave before appearing on TV. Women should limit their use of makeup.

Visuals for Teleconferencing

- Find out ahead of time if and what kind of visuals can be used: document camera, Internet connection, video.
- If you prepare paper visuals for use with a document camera, your documents will fit best if they are laid out in a horizontal, landscape format.
- When you prepare visuals for a videoconference, make the text and images very large and bold or they won't show up.
- To avoid glare from paper visuals, print them on a pastel paper, such as a cream or light blue.

between connecting your teleconference using ISDN or over the company's data network using TCP/IP.

ISDN is a fast digital phone line that has been in use for over a decade and can transfer data at speeds up to 128 Kbps (Kilobytes per second). Videoconferencing systems can use one or more ISDN lines. The advantage of ISDN is that the connection is very reliable. The disadvantage of ISDN is that 128 Kbps is not enough for really high-quality video.

Making videoconferencing calls with TCP/IP networks is useful if your company has already invested in a broadband data network to link offices. Assuming the data network has enough capacity between facilities, you should be able to place high-quality video calls over the data network. The only real disadvantage is that a TCP/IP network can become congested (causing the picture to hiccup or break up). There are several advantages to using TCP/IP. First, you get toll-free calling because your call doesn't go through the public telephone network, so there are no per-minute charges. Second, you get broadcast quality video because the system can run at speeds of 768 Kbps or faster.

Most vendors now sell systems that can talk to both types of networks, so you can choose which method to use on a case-by-case basis. The cost of these systems has also declined dramatically in the past few years so that an entry-level system from Polycom (the largest company in the videoteleconferencing business) now costs less than $2,000.

Many organizations that do teleconferencing have a person who is designated the system operator. This is the person to whom you should address specific questions about the system and how it works.

- If you plan to write on paper visuals, use pastel paper and a dark, heavy marker. Ballpoint pens and pencils don't show up very well.
- If you are using PowerPoint, print out a set of your slides so that they can be used on the document camera in case there are computer problems.
- Talk to the teleconference operator ahead of time if you want to be able to show a DVD or VHS video as part of your teleconference.

Where to Look and Where to Talk

Talking to a camera is different from talking to a live audience. In live presentations, you are told to scan your audience and to try to make eye contact with them. In a teleconference:

- If you are in a room alone, ignore any monitor that shows you. Look directly at the camera as often as possible.

- If you are in a room with an audience in front of you, talk to the audience, but don't completely ignore the camera.
- If there is an audience sitting behind you, talk to the camera most of the time. Turn to look at the people sitting behind you once in a while, but talk to the microphone, even if that means keeping your back to them.
- If you are making a group presentation, adjust your chairs (if possible) so that all presenters are sitting close together and at approximately the same height when seated.
- In a group presentation, talk to the camera or to the last person who spoke.
- Remember that you are sitting in front of an open microphone until the teleconference is over.

How to Behave on Camera
- Fidgeting and squirming will be picked up by the camera. Tapping on the desk, table or podium will be picked up by your microphone and distract or annoy your audience.
- Try to remain calm. Take a deep breath.
- Don't look from side to side.
- Don't look at your watch.
- Keep your hand gestures inside the "box"—between your shoulders and no lower than your armpits.
- Don't interrupt. Raise your hand or gesture to indicate that you would like to speak next.

Starting a Videoteleconference
- Set ground rules and explain the agenda. Prepare agenda in advance and make sure everyone has (or can see) a copy before you begin.
- Establish your protocol with your audience at the beginning of your presentation. Let them know how you want to take questions; is it okay for them to jump in at any time, or do you want them to wait until the Q & A portion of the presentation. Mention any other dos and don'ts that you feel are important.
- Explain to audience members how to use the microphones in the room and show them where the audience cameras are located, so they can make eye contact with the far site when speaking.
- Welcome your audience and thank them for attending.

Concluding a Videoconference
- Thank everyone for participating
- Review any "Next Steps" that were agreed to during the teleconference.
- Turn off microphones.
- Let the system operator know that he or she can close the connection.

CONCLUSION

To speak much is one thing, to speak well is another.
Sophocles, c. 408 BCE

This book started with a discussion of how much speech-making has changed in recent years and how presenters have had to adapt:

- The growing importance of audience analysis and changes in audience expectations
- The way new technology has increased the speed of communications
- The increased importance of visual imagery
- The fact that a much larger group of better educated people need to become effective presenters and to communicate clearly and concisely

To face these challenges, you now have a solid set of guidelines to help you cope with all of these changes and to make you a better speaker. While some people take to giving speeches more naturally than do others, you can become a credible, persuasive professional if you follow the guidelines presented in this book. In their simplest form, the "3Ps" of good presenters are still:

- Prepare carefully.
- Practice often.
- Perform with enthusiasm.

Stick to the 3Ps and you will become a confident presenter—able to to speak competently at school, on the job, or in your community—wherever you may find yourself. As your confidence grows, you will become better at communicating your ideas and you will go further in life. And that's the whole point of learning to do technical presentations and professional speaking.

Appendix: Resources for Presenters

The transient nature of the Internet makes it impossible to know whether websites will be available or usable when you need them. Therefore, instead of using this list as a bibliography, consider it a guide to what you may be able to find online. Note that some of the websites cited are maintained by businesses interested in attracting paying customers. Your best bet for researching any of these topics is to use the keywords given for browsing the Internet and to find articles and books in the library or bookstore.

General Presentation Tips

Keywords: public speaking, presentations, technical presentations, oral communication

Books

Anholt, Robert R. H. *Dazzle 'em with Style: The Art of Oral Scientific Presentation.* New York: W. H. Freeman, 1994.

Axtell, Roger E. *Do's and Taboos of Public Speaking: How to Get Those Butterflies Flying in Formation.* New York: John Wiley & Sons, 1992.

Brody, Marjorie. *Speaking Your Way to the Top: Making Powerful Business Presentations.* Needham Heights, MA: Allyn & Bacon, 1998.

D'Arcy, Jan. *Technically Speaking.* New York: AMACOM, 1992.

Kline, John A. *Speaking Effectively: Achieving Excellence in Presentations.* Upper Saddle River, NJ: Prentice Hall, 2004.

Sullivan, Richard L. and Jerry L. Wircenski. *Technical Presentation Workbook: Winning Strategies for Effective Public Speaking,* 2nd ed. New York: ASME Press, 2002.

Wilder, Claudyne. *The Presentations Kit: 10 Steps for Selling Your Ideas* (rev. ed.). New York: John Wiley & Sons, 1994. See also the author's website *http://www.wilder.com.*

Periodicals About Presentations

These periodicals (printed and online) cover many different aspects of oral presentations. Some focus on the use of technology, while others have more on human factors.

- *Presentations* exists as both a paper magazine and online at *http://www. presentations.com*. It has articles on industry trends, new-product reviews, best presentation practices and how-tos. *Presentations* is for individuals (and organizations) that create and deliver presentations. Subscriptions to the magazine are free.
- *PRISM* is published monthly by the American Society for Engineering Education (ASEE). It focuses on engineering education in the United States. A paper version is mailed to ASEE members. The current issue is posted online for ASEE members only and is password protected. Back issues are available to the public online at *http://www.asee.org/about/ publications/index.cfm*.
- *Technical Communication* is a paper magazine sent to members of the Society for Technical Communication (STC). Effective January 10, 2005, *Technical Communication Online* is no longer available free to those who are not members. Contents of past issues and abstracts are posted online at *http://www.ingentaconnect.com/content/stc/tc*, but if you want the full text of articles, you must pay a fee of $10 per article.
- *Intercom* is published ten times per year and is free with STC membership. You can find information about joining STC at *http://www.stc.org/join.asp*. Subscriptions are not available for non-members. Articles cover new tools and technologies for technical communicators. *Intercom* provides practical examples and applications of technical communication, STC news, columns edited by technical communication experts, and a calendar of industry-related events. You can learn more about the magazine online at *http://www.stc.org/intercom*.
- *T.H.E. Journal* is mailed free to educators in the United States and it can also be found online at *http://www.thejournal.com*. (T.H.E. stands for Technology in Higher Education.) Articles focus on education technology, computer products and high-tech gadgets.

Other Online Resources

Both commercial and non-profit organizations maintain websites related to public speaking. Many offer subscriptions to e-mail newsletters. Others are promoting books or public speaking coaching programs.

- The Advanced Public Speaking Institute website at *http://www.public-speaking.org* has some helpful articles on presentations (examples: audience, organization, audiovisual and props, performance techniques, practice, room set up, handouts) and they will send you their free "Great Speaking Ezine" twice monthly as an e-mail.

- Tom Antion and Associates Communication Company maintains a website at *http://www.antion.com* called "Public Speaking Tips" that has quite a few useful articles.
- The Presenters University website at *http://www.presentersuniversity.com* has many helpful resources for presenters, including free presentation software, free PowerPoint templates, and hundreds of articles. Their free e-newsletter "Presentation Pointers!" is distributed each month.
- Lenny Laskowski is an international professional speaker, author, and consultant who maintains a website at *http://www.ljlseminars.com* with articles that focus on presentation skills. He will send you his free monthly "Simply Speaking" E-Zine with tips about public speaking and presentation skills.
- Consultant Kirby Tepper maintains a website for his Charisma-Consultants group at *http://www.powerpublicspeaking.com*. Tepper frequently adds articles about presentation skills to the content of the site (example: acting techniques for businesspeople). In addition to being a public speaking coach, he also sells copies of his downloadable e-book, *End Stage Fright Now*.

Learning More About Companies and Organizations

As part of your audience analysis, you may need to learn more about a particular company. You may be able to discover more about the company's mission and products by visiting their website. To get a better sense of the corporate culture and to find out what kind of people work there, start at your library. Information about American corporations can be found in *Standard & Poor's Register of Corporations, Directors, and Executives* or the *Thomas Register of American Manufacturers*. Thomas's is available free online at *http://www.thomasregister.com*, but you will have to sign up. Your librarian can direct you to databases where you can find articles about the company and its activities. For more information about a professional association or not-for-profit organization, try the *Encyclopædia of Associations*. To access the online version of this encyclopædia (called *Associations Unlimited*), you will have to go to a library.

Establishing Credibility

Keywords: dress for success, credibility, ethos

Online Resources

The University of Wisconsin–Madison Engineering Career Services has a list of things to consider when you put together a professional wardrobe. It can be found online at *http://ecs.engr.wisc.edu/student/samples/dress.pdf*.

Presenting to International Audiences

Keywords: international business etiquette, intercultural communication

Books

Axtell, Roger E. *Do's and Taboos Around the World.* New York: John Wiley & Sons, 1992.

Morrison, Terri, Wayne A. Conaway, and George A. Borden. *Kiss, Bow, or Shake Hands.* Holbrook, MA: Bob Adams, Inc., 1994.

Morrison, Terri, Wayne A. Conaway, and Joseph J. Douress. *Dun & Bradstreet's Guide to Doing Business Around the World.* Paramus, NJ: Prentice Hall Press, 2001.

Online Resources

CountryWatch provides country-specific geopolitical intelligence on 192 countries on its website at *http://www.countrywatch.com*, including information on each country's cultural etiquette and its economic, political, environmental, and investment situation. The cultural etiquette section is part of the "Social Review" page.

Speakers' Associations

Not all of these organizations are suitable for beginners. Visiting their websites and reading about the mission and membership of the organization will make it easier to decide whether it makes sense for you to join. Most of them charge a membership fee, so it might be wise to ask if you can attend a meeting as a guest before you join.

- Toastmasters International at *http://www.toastmasters.org* has chapters all around the world and encourages new ones to be formed where needed. These clubs meet regularly to give members the opportunity to develop and practice their speaking skills in a supportive environment in which everyone wants to improve. Consult your phone book or check the Internet to contact local Toastmaster chapters.
- Society for Technical Communication (STC) at *http://www.stc.org.* is an individual membership organization dedicated to advancing the arts and sciences of technical communication. It has 25,000 members, including technical writers and editors, content developers, documentation specialists, technical illustrators, instructional designers, academics, information architects, usability and human factors professionals, visual designers, Web designers and developers, and translators. STC has chapters all over the world. At monthly chapter meetings, STC members and invited guest speakers give educational and informative presentations related to communication skills.

- Professional Speakers Guild at *www.professionalspeakersguild.com* is an international organization devoted to helping speakers, trainers, authors, and consultants polish their skills and market themselves.
- International Association of Business Communicators (IABC) at *http://www.iabc.com* calls itself an international knowledge network for professionals in strategic business communication management.
- National Communication Association (NCA) at *http://www.natcom.org* is a non-profit organization of academic researchers, educators, students, and practitioners, whose interests span all forms of human communication.

Persuasive Presentations

Keywords: persuasion, social judgment theory, argumentation, rhetoric, logic, common topics, counterargument, latitude of acceptance

Books

Benjamin, James. *Principles, Elements, and Types of Persuasion.* Belmont, CA: Wadsworth Publishing, 1997.

Breaden, Barbara L. *Speaking to Persuade.* Fort Worth, TX: Harcourt Brace College, 1996.

O'Keefe, Daniel J. *Persuasion: Theory and Research,* 2nd ed. Thousand Oaks, CA: Sage Publications, 2002.

Online Resource

David Straker explains "Social Judgment Theory" at *http://changingminds.org/explanations/theories/social_judgment.htm.*

Impromptu Speaking

Hamilton, Cheryl. *Successful Public Speaking.* Belmont, CA: Wadsworth Publishing, 1996.

Using Your Voice and Nonverbal Communications

Online Resource

The Great Voice Company website at *http://www.greatvoice.com* is the creation of Susan Berkley, the author of *Speak to Influence: How to Unlock the Hidden Power of Your Voice* (Englewood Cliffs, NJ: Campbell Hall Press, 1999). The site promotes Berkley's book and offers information on improving the sound of your voice.

Finding Local Resources

For serious speech problems, the American Speech-Language-Hearing Association (ASHA) certifies speech pathologists. Look in the Yellow Pages of your phone book under "Speech & Language Pathologists" or go to the ASHA website at *http://www.asha.org*.

Coping with Anxiety and Stage Fright

Keywords: anxiety, nervousness, stage fright, social anxiety disorder, self-confidence

Online Resources

- Sandra Zimmer maintains a website at *http://TransformStageFright.com/articles.html*. She offers a free consultation, but there are fees for extended coaching sessions. Zimmer offers classes, workshops, and coaching on Acting for Non-Actors, Speaking from the Heart Presentation Skills, Speaking Voice, Communication, and Executive Presentation.
- Charles di Cagno of the Social Anxiety Center of New York, has a website at *http://www.speakeeezi.com* that includes useful information about Social Anxiety Disorder (of which stage fright can be a symptom) and a "Links Page" featuring dozens of websites related to anxiety, shyness, and fear of public speaking.
- Microsoft offers a huge variety of information about using PowerPoint on its website at *http://www.microsoft.com*. Click on **Office** under the **Product Families** menu, then select **PowerPoint** on the list of products. Many interesting items are listed on the **PowerPoint** page under Powerpoint Highlight, including articles on Grow your skills: Working with pictures, Insert a slide master or title master, Record a sound or voice comment on a single slide, Create a presentation using a design template, How do I use a PDF file in a presentation?, About copying and slide design, and Templates.
- Power Pointers at *http://www.powerpointers.com* has information about how to use PowerPoint visuals effectively.
- PowerPoint Answers *http://www.powerpointanswers.com* is a site maintained by consultants Kathryn and Bruce Jacobs. It has a large collection of articles on using PowerPoint including PowerPoint Tricks, Features and Basics, Issues and Opinions, Reviews, Tools, Template Sources, Missing Features, Sharing Presentations, Public Speaking Hints.

Alternatives to PowerPoint Online Resources

For Windows computers:

- CRE:8 Multimedia can be found on the website at *http://www.presentware.com/products.htm*.
- SmartDraw software is described on the website at *http://www.smartdraw.com/specials/presentation.asp*.

- Impact Engine has a website at *http://www.impactengine.com*.
- Corel Presentations for making multimedia slide shows is part of a software package called WordPerfect Office, found online at *http://www.corel.com*.

For Macintosh computers:

- AppleWorks comes with some Apple computers. It is available from Apple and described on their website at *http://store.apple.com*.
- Keynote is part of a software package called iWork. Learn more from the website at *http://www.apple.com/keynote/*.
- OmniGraffle can be found at *http://store.omnigroup.com*.

Examples of Speeches

Most of the speeches available in books and online are political. Many websites offer the text of great speeches, but a few also have audio or video versions of famous speeches. Your library may have CDs or videos with collections of speeches.

- The History Channel offers audio recordings of "the words that changed the world" at *http://www.historychannel.com/speeches*. The collection is drawn from famous broadcasts and recordings of the twentieth century. To hear the recordings, you must have RealPlayer installed.
- The National Endowment for the Humanities and Michigan State University sponsor a website called "History and Politics Out Loud," which is a searchable archive of politically significant audio recordings at *http://www.hpol.org/*.
- Great American Speeches is a website maintained by the Public Broadcasting System at *http://www.pbs.org/greatspeeches/timeline*. In the speech archives you will find a comprehensive online collection of speech texts from contemporary American history. Here you can read the speeches and backgrounds of many of the most influential and poignant speakers of the recorded age with a brief timeline of historical events.
- C-SPAN offers videos on its website "Historic Ads & Speeches" at *http://www.c-span.org/classroom/govt/video.asp*. Topics include Congress, Defense/Security, Domestic/Social, Economy/Fiscal, International, Judiciary/Courts, Media/Press, Politics/Elections, Science/Technology, State/Local, and White House/Executive. The site is sponsored by the National Cable Satellite Corporation.

Index